BASIC BIOPHYSICS *for* BIOLOGY

Edward K. Yeargers
Georgia Institute of Technology
Atlanta, Georgia

CRC Press
Boca Raton Ann Arbor London Tokyo

Library of Congress Cataloging-in-Publication Data

Yeargers, Edward K.
Basic biophysics for biology / Edward K. Yeargers.
p. cm.
Includes bibliographical references and index.
ISBN 0-8493-4424-7
1. Biophysics. I. Title.
QH505.Y43 1992
574.19'1--dc20 92-7157
CIP

This book represents information obtained from authentic and highly regarded sources. Reprinted material is quoted with permission, and sources are indicated. A wide variety of references are listed. Every reasonable effort has been made to give reliable data and information, but the authors and the publisher cannot assume responsibility for the validity of all materials or for the consequences of their use.

Direct all inquiries to CRC Press, Inc., 2000 Corporate Blvd., N.W., Boca Raton, Florida, 33431.

International Standard Book Number 0-8493-4424-7

Library of Congress Card Number 92-7157

Printed in the United States of America 1 2 3 4 5 6 7 8 9 0

Printed on acid-free paper

PREFACE

The revolution in physics that followed World War I was paralleled by the revolution in biology that followed World War II. The latter was spearheaded by physicists and chemists who found fertile ground for change in the descriptive approach then prevalent in the life sciences. The period since then has been marked by a dramatic shift in emphasis from description to explanation, and most biologists now must study physics and chemistry as a routine part of their education.

The quest for better explanations in biology has relentlessly led to the study of molecular structure and thermodynamics. One thing that has been made clear as a result, at least to anyone who has taken a little trouble to study the subject, is that living systems are subject to the same rules of physics that nonliving systems are. The unique structures of biological molecules and the not-at-equilibrium behavior of biological processes contradict no known physical laws.

I have included here explanations of most of the biophysical phenomena that are introduced in that part of a one-year introductory biology course which is devoted to thermodynamics and to molecular structure and which reappear constantly in advanced courses in modern biology. Equally important, this means that I have not covered a number of "traditional" biophysical subjects, e.g., hearing and sedimentation.

This is not a popularized account; much of the material is complex and will require considerable effort on the student's part. A few of my explanations are somewhat rigorous, but most rely on plausibility more than on rigor. When I felt that detailed explanation would take the narrative unnecessarily far afield, I adopted a please-take-my-word-for-it approach and included suggestions for extra reading. In fact, the sections entitled "Applications, Further Discussion, and Additional Reading" at the end of each chapter are integral parts of the chapters.

To back up the narration and to compensate for the absence of quantitative descriptions, I have included many illustrations, a feature that appeals to students of biology. The problem with illustrations, however, is that they convey the impression that the object pictured actually looks ***that way***, although maybe smaller, and that the only problem is one of magnification. Biology students like things to be visually realistic and so are a bit susceptible to this misconception. As a result, I have made a major point throughout the book of the distinctions between macroscopic "reality", models, and submicroscopic phenomena.

There are few equations in this book because my intended readership will not be performing many calculations. Just one chapter is genuinely quantitative and it involves algebra only — that is Chapter 18, describing the absorption spectrophotometer. This instrument is frequently used by students of biology and biophysics, but they sometimes feel that it is little more than a mass of dials and knobs. The usefulness of the instrument is greatly multiplied, and the possibility of error greatly diminished, by some knowledge of its internal mechanics. I felt that a chapter on its use and function, right after the chapters on molecular energy levels, would be appropriate.

Another instrument frequently found in biology laboratories is the microscope. I have not discussed its workings, only the concept of resolution, which helps to clarify the relationship between the thing observed and the means used to observe it.

It has been my good fortune for some time to teach several biology courses per year to a mixture of majors from among Georgia Tech's bright undergraduates. This text grew out of the needs of those courses. The actual decision to put pen to paper came about because the Georgia Tech Foundation and the Botany Department of The Hebrew University of Jerusalem were kind enough to give me the chance to stare out of my office window during a sabbatical and thereby to ruminate on everything under the shining sun of Israel, which is a lot.

I appreciate the help of several people: Drs. Jaroslav Drobnik, David Dusenbery, Petr Hochmann, Albert Mayer, Alexandra Poljakoff-Mayber, Lenora Reinhold, Thomas Tornabene, and Turgay Uzer. I am grateful to Christopher, Fran, Hugh, Jon, and Pauline for being there when it got dark.

The Author

Edward K. Yeargers is associate professor of biology at the Georgia Institute of Technology. Dr. Yeargers received a B.S. degree in physics from Georgia Tech, a M.S. degree in biology from Emory University, and a Ph.D. degree in biophysics from Michigan State University. He has served postdoctoral and sabbatical appointments in radiation physics at Oak Ridge National Laboratory, Oak Ridge, Tennessee, in theoretical chemistry at the Institute of Physical Chemistry in Prague, Czechoslovakia, and in plant physiology at The Hebrew University in Jerusalem, Israel. He is a member of the American Institute of Biological Sciences and The Biophysical Society. Dr. Yeargers has published research papers in radiation biology, molecular spectroscopy, and molecular structure, but he especially enjoys teaching.

CONTENTS

Chapter 1
The Use of Models in Science

Biophysics is the Scientific Application of Physics to Biology

The intent of this book is to describe some physical phenomena that are relevant to the study of biology. This kind of application of physics to other scientific fields is so widespread that whether people call themselves, say, physical chemists or molecular biophysicists sometimes boils down only to which academic department awarded them their degree.

Models Simplify Scientific Life

A biologist might be interested in a particular molecule, a beaker of solution, a cell, or a deciduous forest. Each of these is so complicated that it would be necessary to represent it by a simplified scheme or structure called a ***model***, having fewer parts and interactions. A model is a representation, or "mental picture", of the actual physical system and is obtained by stripping the actual system of all behavior except that which bears directly on the problem. The model can then be tested to see if it predicts any of the properties of the "real" system that it represents.

Of course, one always runs the risk of discarding important factors under the guise of simplication. This especially leads to serious problems if initial results seem to vindicate that simplification; the agreement between a model and an actual structure frequently proves to be fortuitous.

Something else we should be careful about is that by definition a model must be described in terms from our everyday, macroscopic, "real" world, and we therefore have a natural tendency to assign a "reality" to our models also. This isn't necessarily bad — in fact it may be the only sensible approach. After all, we have only the model to work with, having decided that the original system was too complicated for direct understanding. However, we must always bear in mind that the model is a ***representation*** and the "reality" we associate with the model is going to change whenever we change the model — which we inevitably will do.

WHAT IF A MODEL IS WRONG?

If the philosophy of the previous three paragraphs bothers you a little, you are in the company of veterans. Any practical approach to science necessitates simplifying assumptions, but it is clear that there is often a fine line between making a problem intellectually tractable on the one hand and losing its essential nature on the other hand.

It may help to allay your uncertainties by bearing in mind that the history of scientific inquiry shows that it is useless to ask about what something "really is". Rather, scientific inquiry generally involves the construction of better and better model representations. Each model is progressively revised, sometimes dramatically, to remain consistent with the outcomes of new experiments suggested by the model itself. Thus, ***in the long haul, virtually all models prove to be incorrect, but even an incorrect model can suggest the way to a better one.*** For instance, the biological model called "Lamarckianism" — the inheritance of acquired characteristics — is incorrect in terms of present knowledge, but that does not mean that it resulted from bad science. In fact, it seemed to answer many questions posed at the time of its inception, e.g., why musicianship sometimes runs in families. It further suggested that an experiment of the sort, say, of cutting the tails off each of a line of mice would eventually result in a line of naturally short-tailed mice. The fact that things did not work out the way the Lamarckian model predicted led biologists to seek other models for inheritance and evolution. The fact that a model eventually proves to be inadequate should not lead us to be scornful about it; indeed, a mark of a good model is that it leads to its own rejection by suggesting experiments whose outcomes are so totally unexpected that a new model becomes necessary.

Applications, Further Discussion, and Additional Reading

1. As an example of modeling, represent a human being as a light bulb. What size bulb (in watts, W) produces heat at the same rate as the human? Assume the bulb loses 90% of its input energy as heat and that the human consumes 2000 cal/day, all of which end up as heat. You should note that a dietitian's calorie is 1 kcal on a physical chemist's scale. (Answer: 106.6 W.)
2. A survivorship curve is a plot of the number of surviving individuals vs. time, assuming zero birth rate. As a model, suppose that 1000 test tubes were purchased for a lab and that 10% of those remaining were broken each month. Thus, the survivors are 1000, 900, 810, etc. Plot a survivorship curve for these data. An elaboration of this model and some analogous

biological data for a cohort of wild birds can be found on page 1039 of *Life*, 3rd ed., by Purves, W. K., Orians, G. H., and Heller, H. C., Sinauer Publishers, Sunderland, MA, 1992.

3. The model of evolution called "Inheritance of Acquired Characteristics" was replaced by the Darwinian model, a basic conclusion of which is that properties favored by selection can be transmitted to offspring. Darwin proposed an utterly incorrect model, the "Provisional Theory of Pangenesis", for the ***mode*** of this transmission, but that in no way detracts from the usefulness of the model of Darwinian evolution itself, the evidence for which is overwhelming and which is discussed in any introductory biology text. A discussion of Darwin's ideas about pangenes, the basic concept of which actually dates back to Aristotle, can be found in *History of Genetics* by Stubbe, H., 2nd ed., 1965, translated by Waters, T. R. W., The MIT Press, Cambridge, MA, 1972, pp. 172–175.
4. There is a discussion of modeling in science in "Is science logical?", by Pease, C. M. and Bull, J. J., *BioScience,* 42(4), 293–298, 1992. (The authors answer "no" to their title question.)

Chapter 2
THE OBSERVATION PROCESS

CAN ANALYSIS EVER BE TRULY OBJECTIVE?

A scientist (or artist) is really an observer/interpreter, taking note of his or her surroundings and then trying to interpret those observations in some greater context. A scientist might see the mating behavior of a nematode in terms of chemical attractants, e.g., pheromones, found in all animals, including humans. On the other hand, an artist might paint a picture of a scene or write a poem about it, depicting his/her own reactions to the scene, and thus capture some universal feature of that scene. There are some interesting relationships between the processes of observation and interpretation. This chapter will examine some consequences of this interdependency.

We often convince ourselves that we are neutral observers and that our observations and interpretations are "objective". Several things suggest that this conviction is overly optimistic. First, we are all victims of prejudice, having an emotional investment in the outcome of all our activities, scientific or not. In fact, experimenters usually try to build elaborate safeguards into their experiments to minimize the effects of personal bias. As an example, when a new vaccine is first tested, some patients must randomly be given the real vaccine and others must be given an inert placebo. As heartless as the latter act may seem, it is necessary to eliminate the possibility that the immunizing effect is due to the *act* of vaccine administration or to something in the carrier fluid for the vaccine. If the participating physicians can guess which solution is the real vaccine, they might administer it — rather than the placebo — to all their patients. Thus, the developers of vaccines go to great lengths to make the actual vaccine and the placebo look alike.

A second problem with our image of ourselves as neutral observers is in the notion that the actual act of watching, touching, or listening to something doesn't influence the nature of that thing. We can take the simple process of observing a hen's egg and show that detached objectivity — perception untainted by the observer — is not really possible, although its importance may be small (as explained below). Under the white light of an ordinary desk lamp, an egg appears white. However, under a red light the egg appears red, and so on. In the dark an egg has no color at all, but we might feel it and say that it is smooth and ovate. Evidently what we perceive the egg to be depends to some degree on the means we use to perceive it.

The egg example demonstrates a very fundamental principle: what something "is" is in large measure dependent on the observation process itself. This has the far-reaching consequence of making "reality" relative, i.e., relative to the observation process. The observer can never be totally detached from the observation process because he or she chooses the means of making the observation, and that in turn determines the nature of the object observed. Someone else might choose a different, but valid, method of observation and thus arrive at a different, but valid, conclusion. The notion of an "absolute reality" shared by everyone, thus becomes much less credible.

A third difficulty with our roles as supposedly objective observers was mentioned in Chapter 1; it is our need to form a "mental picture", drawn from everyday life, of a phenomenon or thing which is ***not*** a part of our everyday life. In molecular biophysics this means that the things being interpreted are at the submicroscopic scale, while our observations and interpretations are at the macroscopic scale. That disparity of scale has unexpected consequences. As an example, we could run an experiment to locate an electron by letting it hit a screen, such as the one on a television. The macroscopic model of such a collision is that of a projectile hitting a target and leaving a mark at the site of the collision: a pulse of light on the screen tells us where the electron hit and therefore where it was at the time of the collision. The phrase, "where it was" — past tense — is crucial because after hitting the screen the electron cannot subsequently be located. It is (perhaps) somewhere in the glass or the scintillator, but we will never know because it cannot be made to hit a second screen. Merely locating that electron a single time has forever cost us the ability to investigate that electron again.

The same problem doesn't occur if we use the same macroscopic model for a collision involving an actual macroscopic object: let a car hit a large paper screen. The hole in the paper tells us where the car was, but doesn't affect the car. We are confident that the car's trajectory is unchanged by the observation; at this level, observation does not appreciably disturb the object observed.

The problem for us of course is that we aren't interested in locating cars — we want to locate electrons, the observation of which requires comparatively large-energy-scale methods which seriously disturb electrons. This disturbance has no analog in our macroscopic world, a world which unfortunately forms the basis for all our models! We thus find ourselves in the peculiar position of constructing models out of a world that does not act like the one from which the biophysical systems originate.

We should note that the disparity-of-scale problem also occurs for extremely large objects. We might fancy that looking at a distant star through a telescope is just like looking through the telescope at a bird 50 ft away — a virtually instantaneous, straight line-of-sight view. Yet, light from stars may have started out billions of years ago and its path can be bent by passage near massive stars.

Applications, Further Discussion, and Additional Reading

1. Alexander Kohn's book, *False Prophets*, is a discussion of fraud in science. Kohn points out the role played by self-interest in causing scientists to make conscious or unconscious errors in taking, analyzing, and reporting data. Even Isaac Newton seems to have done it! (Kohn, A., *False Prophets*, Basil Blackwell, Cambridge, MA, rev. ed., 1988.)
2. The thought that certain features of our physical world may not be accessible bothers many people. If the topic of "impossibility" interests you, you may want to read *No Way: The Nature of the Impossible*, Davis, P. J. and Park, D., Eds., W. H. Freeman, San Francisco, 1987. This book contains an article written by Park, entitled "When Nature Says No".

Chapter 3
ELECTROMAGNETIC RADIATION — A GENERAL DISCUSSION

ELECTRIC CHARGES AND THEIR INTERACTIONS

The words "positive" and "negative", applied to electric charges, arise out of the need to combine aggregates of charge arithmetically and thus to obtain their net charge: ten electrons (negative charge) and nine protons (positive charge) have a net charge of −1. This nomenclature has the unfortunate side effect of suggesting that electrons electrically lack something that is present in protons. It will not do to suggest that the quality lacked is "positiveness" because we would then have to say that protons lack "negativeness". In fact, there is no reason why the original assignment of charge sign could not have made electrons "positive". The point of this discussion of semantics is to emphasize that the charges on electrons and protons have a perfect mirror symmetry, differing in sign only, their absolute magnitudes being identical. Differences in other aspects, such as mass, are irrelevant to their electrical properties.

The fact that most material is electrically neutral masks the fact that it is composed of large quantities of positively and negatively charged particles. If there is an imbalance of those

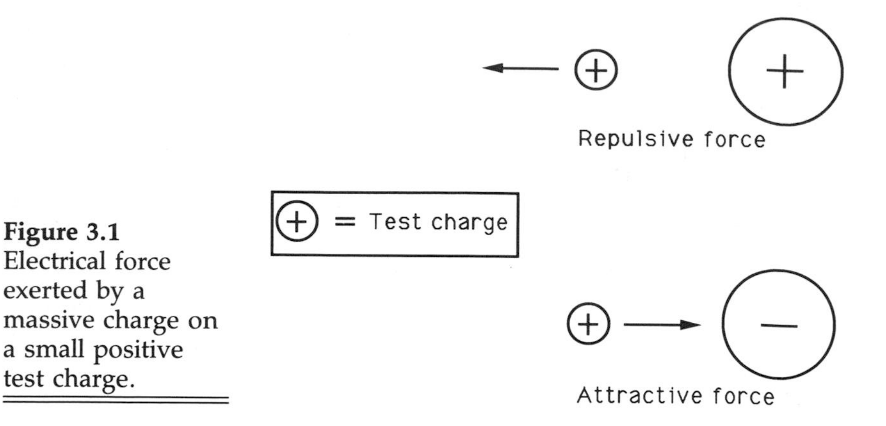

Figure 3.1 Electrical force exerted by a massive charge on a small positive test charge.

charges in an object, the object is said to be electrically charged. It is an empirical fact that two electrically charged objects exert a force on one another. The interaction of charged objects can be distilled to the statement, "unlike charges attract; like charges repel."

We can study the interaction of two electrically charged objects in the following way: suppose that, as in Figure 3.1, there exist a massive charge and a small positive charge, the latter being called a ***test charge***. Depending on its sign, the massive charge will exert a repelling or attracting force on the test charge, as shown by the arrows in the figure. Of course, the test charge will exert an equal force on the more massive one, but the latter will remain stationary because of its mass. In practice the force on a test charge can be measured by suspending it with a thin filament and measuring how much it is displaced by the larger charge.

Physicists say that an ***electric field*** exists wherever there is an electric force on a positive test charge; the direction of the field is the same as the direction of the force. Thus, the arrows in Figure 3.1 and subsequent figures in this

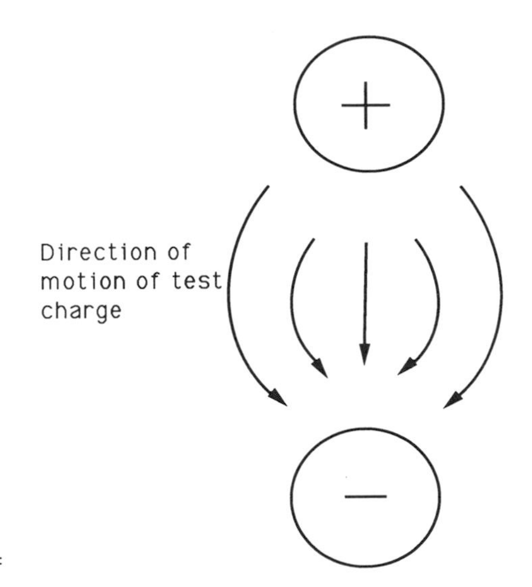

Figure 3.2
Electrical forces near an electric dipole. The test charge itself is not shown.

chapter serve to define not only the direction of that force, but also the direction of the electric field.

We can now replace the larger charge by a pair of charges having equal magnitude but opposite signs. Such an arrangement is called an ***electric dipole***, as shown in Figure 3.2. If the two charges in the dipole remain fixed with respect to one another, as is the case here, the dipole is said to be stationary. A test charge, not shown in the figure, will be attracted to the negative pole and repelled by the positive pole, as indicated by the arrows.

Electromagnetic Radiation and Oscillating Dipoles

Suppose now that the dipolar charges move up and down, opposite to one another, along their connecting axis, as shown in Figure 3.3. This dipole is said to oscillate — in contrast to the stationary dipole of the previous figure. Devices to create this motion of charges are called oscillators and are well known to electrical engineers; the details need not concern us here. We need merely note that one of the charges moves upward, decelerates, stops, and

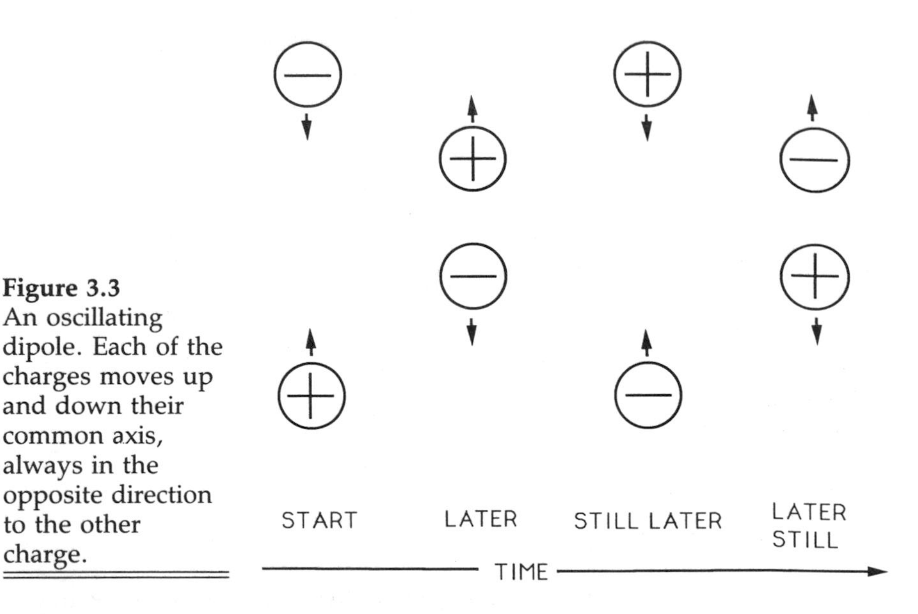

Figure 3.3
An oscillating dipole. Each of the charges moves up and down their common axis, always in the opposite direction to the other charge.

then accelerates in the other direction. The second charge behaves the same way, but in the opposite direction to the first one. The two charges are distinct from one another except when they pass each other at the center. The structure containing the moving charges, and defining the axis of motion, is called an ***antenna***, or dipole antenna.

Suppose a dipole antenna has charges that ***oscillate*** at a constant frequency. The electric force generated near the antenna will thus vary with time, depending on the relative orientations of the two charges in the dipole. For example, if the dipole were positive at the bottom, a nearby test charge would experience a force upward toward the negative pole; if the dipole were positive at the top, a nearby test charge would experience a force downward toward the negative pole; if the positive and negative dipolar charges were in the center of the dipole, a nearby test charge would experience no force at all.

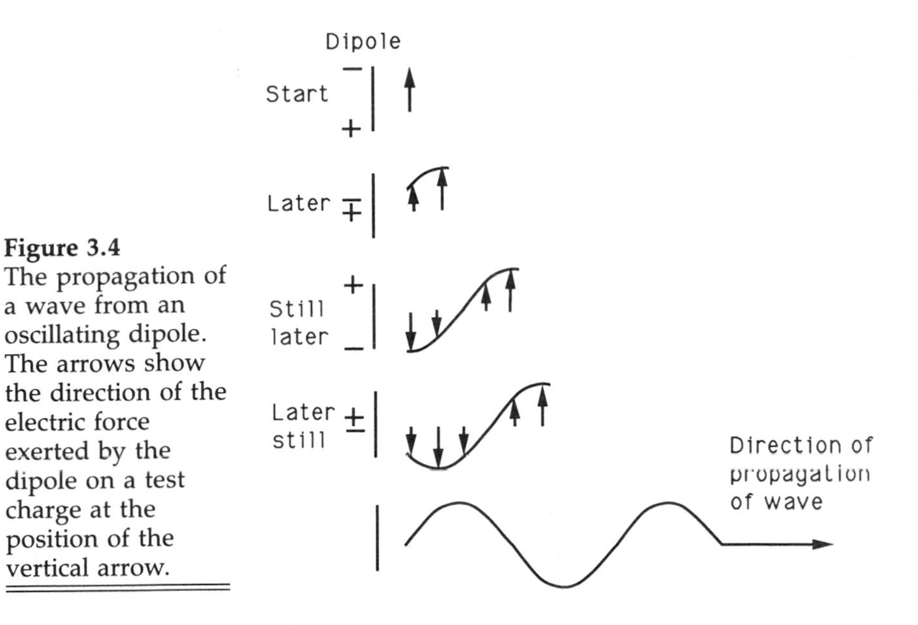

Figure 3.4
The propagation of a wave from an oscillating dipole. The arrows show the direction of the electric force exerted by the dipole on a test charge at the position of the vertical arrow.

Suppose, however, that the test charge were ***not*** near the oscillating dipole, but rather at some distance away. In that case the changes in the dipole's charge orientation would not be felt instantaneously by the test charge. Instead, those changes would be propagated from the antenna to the test charge at the speed of light (c), $c = 3 \times 10^{10}$ cm/sec. Figure 3.4 shows the propagation of an oscillating electric field toward the right from a dipole antenna. This oscillating electric field in space is called ***electromagnetic radiation*** and we see that it merely represents an electrical disturbance propagated away from an oscillating dipole.

Electromagnetic radiation is generated by the periodic acceleration and deceleration of a dipole's charges. These changes in velocity occur when the dipolar charges reach the end of the antenna, slow down, stop, and then go the other way. Thus, a static electric dipole would not produce such radiation. A varying magnetic field accompanies the varying electric field from a dipole, but the magnetic field does not affect anything of interest to us, so we will ignore it.

Several parameters are used to describe electromagnetic radiation. The amplitude is the height of the wave. Examination of Figure 3.4 reveals that the amplitude alternates in sign and direction at different times and positions, coinciding with the up-and-down motion of the charges of the dipole. The wavelength, λ, of the radiation is the peak-to-peak distance between waves. The energy, E, of the radiation is related to the wavelength by $E = hc/\lambda$, where h is Planck's constant, 4×10^{-15} eV·sec. We will henceforth use electron volts (eV) as our energy unit, where 1 eV is the energy given to an electron when it moves through a potential of 1 V. This is equivalent to saying that the electron starts out at the negative pole of a 1-V battery and is accelerated to the positive pole; this gives the electron 1 eV of energy.

There Are Many Sources of Electromagnetic Radiation

The antenna of Figures 3.3 and 3.4 is macroscopic. At the submicroscopic scale the antenna is replaced by an emission source at that level. As usual, it is not possible to examine such a submicroscopic structure directly and we must be content to think of it as acting like some kind of antenna. Thus, no matter what the scale the source will be some representation of an oscillating dipole (or similar structure). The next table describes several kinds of electromagnetic radiation. Even though the members of the list have different names and are created at different levels of atomic and molecular organization, remember that they all stem from accelerated electrical charges and that they all move at speed c if they exist at all.

Some kinds of electromagnetic radiation are

1. γ-Radiation — Originates in changes in radioactive nuclei; typical energies are 0.5 meV to several meV (10^{+6} eV).
2. X-Radiation — Originates in changes in the energy states of electrons of inner-lying electrons of heavy atoms; typical energies are 50 to 300 keV (10^{+3} eV).
3. Ultraviolet (UV) light — Originates in changes in the energy states of outer electrons of most atoms; typical energies are 5 to 10 eV.
4. Visible light — Same origin as UV; different wavelengths are perceived by our eyes as different colors; typical energies are 2 to 5 eV.
5. Infrared (IR) radiation — Originates in changes in the energy states of vibrations of two nuclei across a chemical bond; perceived as heat; typical energies are 0.1 to 0.7 eV.

This list is intentionally brief and approximate; there are many other kinds and sources of electromagnetic radiation. The kinds of electromagnetic radiation listed above are each discussed in more detail in subsequent chapters.

Applications, Further Discussion, and Additional Reading

1. A discussion of the general nature of electromagnetic radiation and its origin in oscillating electric dipoles can be found in any introductory physics text. For example, a readable, calculus-based treatment is found in *University Physics*, 5th ed., by Sears, F. W., Zemansky, M. W., and Young, H. D., Addison-Wesley, Reading, MA, 1980, pp. 627–652.

Chapter 4
Electromagnetic Radiation — A Wave

Criteria for the Wave Description

In Chapter 3 we saw how an oscillating dipole could generate electromagnetic radiation. In this chapter we will see how the wave nature of electromagnetic radiation can be justified, even when the nature of the source (antenna) cannot be described.

Our instruments and we are macroscopic, so we must use a macroscopic criterion to decide whether or not a physical entity behaves like a wave — even though that entity is at the submicroscopic level. In the following discussion we will examine some properities of macroscopic-world waves, namely, how they can reinforce or cancel one another. We will then show how certain submicroscopic-world phenomena also show reinforcement and cancellation, thus justifying a wave description of the submicroscopic phenomena. For reference, Figure 4.1 graphically shows the meaning of the wavelength (peak-to-peak distance) and amplitude (height) of a wave. From this point on we will say that two electromagnetic waves are identical if their maximum amplitudes and wavelengths are the same.

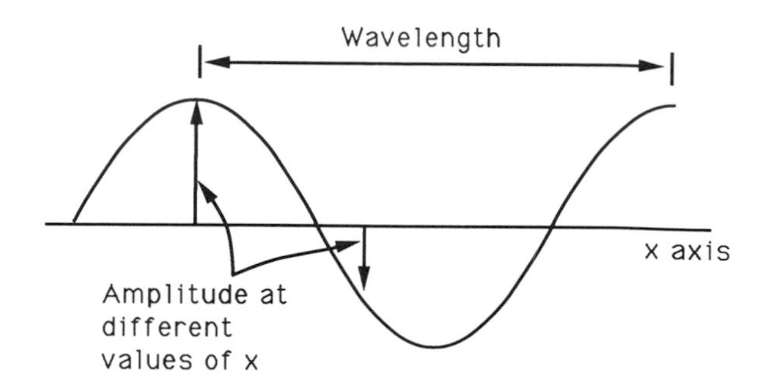

Figure 4.1
Definition of parameters for a wave.

A well-known wave property is that of interference. Figure 4.2 shows the arithmetic sum of two identical waves when their peaks meet and their troughs meet. They are said to be ***interfering constructively*** because the resultant wave is bigger (more intense) than either of the two contributing waves.

Figure 4.3 shows that the arithmetic sum of two identical waves is zero when the peaks of one coincide with the troughs of the other. This condition is called ***destructive interference.***

In a more down-to-earth situation, suppose that two people on opposite sides of a body of water each make a single splash at about the same time and we follow the two sets of waves as they move toward one another at the center of the body of water. Careful observation would show the situation in Figures 4.2 and 4.3. As the two sets of waves approach

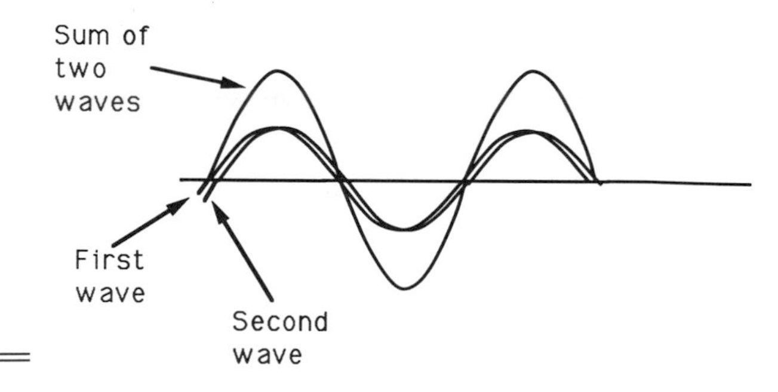

Figure 4.2
Addition of two waves in phase.

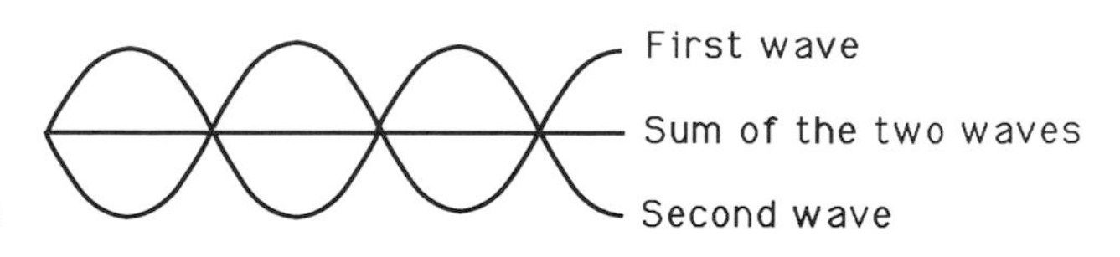

Figure 4.3
Addition of two waves completely out of phase.

and pass one another, their amplitudes combine to give, alternately, waves with double the individual amplitudes and then zero amplitude. Again we have constructive and destructive interference, respectively.

The two descriptions above are typical of wave behavior, not of particles, and we can henceforth use the phenomenon of interference as a defining property of waves. Always bear in mind, though, that it is a criterion from the macroscopic world.

We Apply the Macroscopic Description to Electromagnetic Waves

Suppose now that the wave is a varying electric field with wavelength 500 nm (1 nm = 10^{-9} m). In other words, some source is radiating this varying electric field, as shown in Figure 4.4. (It was mentioned earlier that there is a magnetic field varying simultaneously with the electric field, but the magnetic field can be ignored for our purposes.)

It is easy to design an experiment demonstrating that the electromagnetic radiation's behavior is consistent with a wave description. As shown in Figure 4.5, we imagine two identical waves starting out from the source with

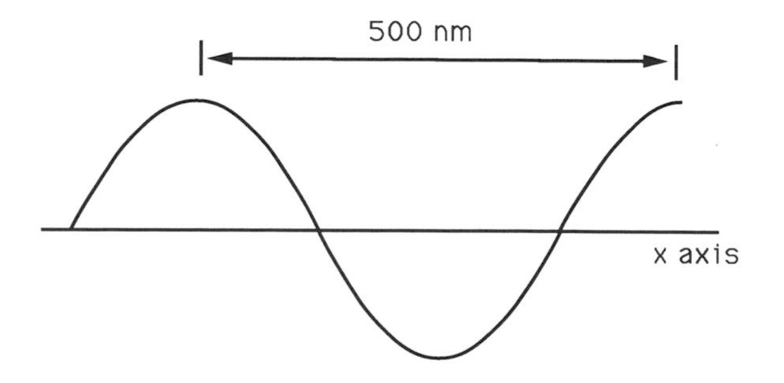

Figure 4.4
A wave with wavelength 500 nm.

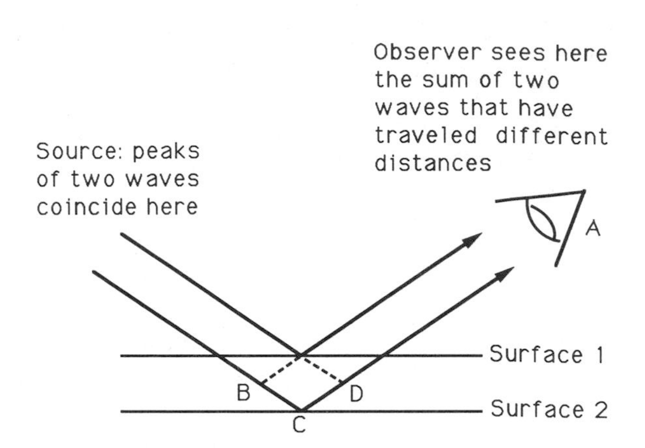

Figure 4.5 Interference between two waves. The two waves start out in phase at the source, but travel different distances to the eye. Their phase relationship at the eye will be determined by the extra distance (BC + CD) that the lower wave must travel.

their respective peaks and troughs aligned (they are ***in phase***), and each is reflected from a different surface. For simplicity the waves are shown as straight lines, with the oscillatory motion omitted.

Thus, the two waves travel different distances. An observer at point A will see the arithmetic sum of the two reflected waves. If the extra path length (BC + CD) is half a wavelength (250 nm), the resultant amplitude will be zero — interference is destructive — because the two waves will be completely out of phase. The situation will be similar to that of Figure 4.3 and no radiation will be observed. On the other hand, if the extra path-length is exactly one wavelength (500 nm), the two waves will be in phase and interference will be constructive. The situation will be similar to that of Figure 4.2. These results are exactly those that are seen experimentally, thus justifying a wave description of electromagnetic radiation.

As mentioned earlier, the energy of an electromagnetic wave is $E = h\nu$, where h is

Planck's constant, 4×10^{15} eV · sec, and ν is the frequency of the wave. Frequency and wavelength, λ, are related by $\nu\lambda = c$. Thus, we can express energy as $E = hc/\lambda$, meaning that energy and wavelength are inversely proportional — when one gets bigger, the other gets smaller.

THE OPTICAL MICROSCOPE AND THE NOTION OF RESOLUTION

All biology students eventually find themselves using an optical microscope. The word "optical" merely means that the electromagnetic radiation used in the device is visible light, i.e., its wavelength can be detected by our eyes. The optical microscope bends the path of the light originating from an object and thereby forms an enlarged image of the object on our retina or on a piece of photographic film. The bending of the light path by a lens is called ***refraction*** and is a wave property.

There is a limit to the useful magnification of any optical device and it is set by several factors. Some of these can be collectively called ***aberrations*** and they are inherent in the structure of all real lens systems. For example, the point at which a simple lens focuses light depends on the wavelength and, therefore, a source of two or more wavelengths of light cannot be focused sharply ("chromatic aberration"). Fortunately, most lens aberrations can be corrected to a considerable degree by the use of multiple, or compound, lenses. The difference between an expensive and a cheap microscope can often be traced to the sophistication with which the various aberrations have been designed out of the device.

It happens that there is a more profound reason than aberrations for the limit on magnification; it is one set by quantum mechanics.

In order to understand this point, we first need to state the basic purpose of a microscope: it is to enhance the distinction between very small objects, e.g., adjacent parts of a chromosome. The microscope is doing its job if it allows us to see two small objects as being ***separate***, rather than as a single blurred unit. Put another way, the microscope must provide us with ***details***; magnification without details is useless.

The smallest separatory distance between two objects that we can visualize with a microscope is called the ***resolution*** and, among other things, it is related to the wavelength of the illuminating light. As an analogy, the accuracy of a meter stick is approximately limited to the space between the stick's finest gradations (perhaps millimeters). By the same token, the resolution of an optical device cannot be better than the smallest gradations of the radiation, namely, the wavelength of the radiation. To a good approximation we can thus say that a microscope cannot resolve two objects whose separation is less than about 400 nm, near the shortest wavelength in visible light. Of course, the resolution could well be ***worse*** than that because of the lens aberrations mentioned earlier.

Students sometimes have the impression that greater magnification is all that is required to visualize small objects, but we see from this discussion that greater magnification would only yield bigger blurs! On the other hand, shorter wavelength radiation ***would*** help, but that would mean using ultraviolet (UV) light — which is absorbed by the proteins of our corneas and eye lenses and which we therefore cannot see. UV light is also absorbed by glass and that would necessitate quartz lenses in the microscope, making the cost prohibitive. The solution lies in the electron microscope, which will be discussed in Chapter 7.

We now can understand why bacteria are so difficult to visualize with ordinary microscopes: their dimensions are close to the wavelength of visible light. For example, a *Salmonella* cell is about 500 nm long and 100 nm wide. Viewing *Salmonella* with visible light is somewhat like trying to measure accurately the width of a pencil lead with a ruler marked in millimeters.

Applications, Further Discussion, and Additional Reading

1. A calculus-based discussion of mechanical wave motion and interference can be found in *University Physics*, 5th ed., by Sears, F. W., Zemansky, M. W., and Young, H. D., Addison-Wesley, Reading, MA, 1980, pp. 363–388.
2. A discussion of the capabilities of various kinds of microscopes, including an X-ray microscope, is found in "Advances in microscopy," by Root, M., *BioScience*, April 1991, pp. 211–214.

Chapter 5
Electromagnetic Radiation — A Particle

Light can be Represented by Indivisible Quanta or Photons

Having just described electromagnetic radiation as a wave, it may seem a bit peculiar to embark on a particle description of that radiation. Perhaps it will help to recall from Chapter 2 that our perception of something depends in part on the means we use to perceive it and that we used a specific experiment (interference) to define the wave property. There is no reason to believe that one human-invented experiment is the only one that can be used to describe electromagnetic radiation. In fact, the ***Einstein photoelectric effect,*** described next, is consistent with light being a particle.

If ultraviolet (UV) light of appropriate energy, $E = h\nu$, illuminates a metal surface, electrons are emitted from that surface. The energy of each electron is E, minus an amount necessary to overcome the metal's tendency to bind the electron. (The metal ***must*** hold the electrons to some degree, otherwise the electrons would drift away!) The important point is that the light energy disappears as a single unit of energy, E, which is what we would expect from an indivisible particle. This is, however, not what we would expect from a wave, the extent of whose exposure could be varied to any degree merely by turning it on and off whenever we chose.

In another famous experiment, A. H. Compton bombarded orbital electrons with X-rays. In these collisions the X-rays and the electrons bounced off one another in a manner very similar to the way billiard balls bounce off one another. We are thus forced to interpret the submicroscopic interaction of X-rays and electrons in terms of a macroscopic-world model in which both the X-rays and the electrons have mass and volume, i.e., both the X-rays and electrons act like particles, at least in this experiment.

We note in passing that the energy of the light particle, or photon, is still expressible as $h\nu$, where h is Planck's constant and ν is the frequency — a wave property. This is no contradiction; it is merely an acknowledgment that the energy of the light must be the same whether we use a particle description or a wave description.

Applications, Further Discussion, and Additional Reading

1. Discussions of the photoelectric effect and the Compton effect can be found in *Fundamentals of Physics*, 2nd ed. extended; by Halliday, D. and Resnick, R., John Wiley & Sons, New York, 1981, pp. 773–786. A second suggested source is *Introduction to Modern Physics*, 2nd ed., by McGervey, J. D., Academic Press, Orlando, FL, 1983, pp. 103–111.

Chapter 6
The Electron — A Particle

The Electron can be Described as a Charged "Point"

If an electron collides with a fluorescent screen, a small pulse of light is emitted by the screen, a result consistent with the electron being a particle. (A wave would have spread, as waves do, and would have illuminated an ***area*** of the screen.)

A second demonstration of an electron's particle-like behavior is shown in a ***Wilson cloud chamber.*** The electron traverses supersaturated water vapor in the chamber, leaving in its wake a line of ionized water molecules. The nearby supersaturated water condenses onto the ions, forming a macroscopic line of water beads which trace the trajectory of the original electron. Only a particle can have a line-like trajectory — a wave would spread out over time. We should note that these criteria are at the macroscopic level, as always.

Applications, Further Discussion, and Additional Reading

1. The Wilson cloud chamber is described in *University Physics,* 5th ed., by Sears, F. W., Zemansky, M. W., and Young, H. D., Addison-Wesley, Reading, MA, 1980, pp. 805–806.

Chapter 7
The Electron — A Wave

The Electron can be Described as an Electrical "Disturbance"

We have already used the concept of interference to show that electromagnetic radiation can be described as a wave and we can use the same concept for electrons. Davisson and Germer performed an experiment, similar to that described in Figure 4.5, in which ***electrons*** were reflected from parallel layers of atoms in a crystal. The electrons showed interference, thus demonstrating the wave nature of the electron.

We should wonder what such a description means. First of all, we must remember that the interference experiment can only show wave properties, if anything, and that no conclusion regarding particle properties can come from such an experiment, ***even if no interference were shown.*** The electron, however, ***does*** show interference, as does all matter, and we feel naturally compelled to place some macroscopic-world interpretation on that. It turns out that doing so is not easy. In any event, don't think of the wave nature of the electron as meaning that the electron bobs up and down, because that implies a discrete object — a particle!

Louis de Broglie postulated a wave model of the electron, in which an electron orbit was represented by a group of in-phase waves. The model predicted that the angular momentum of an electron could take on only a limited number of values, a concept that Neils Bohr had introduced some years earlier as an *ad hoc* assumption. This had led Bohr to the concept of the planetary atom which, as we will see in Chapter 10, yielded many correct predictions of electron behavior. Thus, de Broglie's wave model was backed up by experiment; de Broglie did not offer any explanation of what these "matter-waves" represented, however.

Probably the best one can do is to acknowledge that there ***is*** some kind of wave associated with the electron, the evidence for which is twofold: interference in an experiment at the macroscopic level and de Broglie's postulation of a wave property previously having been shown to lead to correct predictions of electronic behavior. Historically, the notion of matter-as-wave led to the formulation of quantum mechanics, also called wave mechanics, in which the "position" of a particle is represented as a wave of probability. Serious consideration of that would take us far afield, but the idea will reappear in our discussion of electron orbitals in Chapter 11.

The Electron Microscope and the Notion of Resolution

In Chapter 5 we saw how the wavelength of visible light placed an intrinsic limit on the resolution attainable from the light microscope, namely, that two points closer than about 400 nm would appear to be a single image and that further magnification would merely increase the apparent size of that single image. Shorter-wavelength electromagnetic radiation such as ultraviolet (UV) requires expensive quartz optics and is dangerous to the eyes. X-rays, which have very short wavelengths, cannot be focussed; for the most part, they pass straight through matter (hence their medical utility). On the other hand, electrons can be given a very wide spectrum of wavelengths and can also be focused; these two properties provide the basis for the electron microscope.

In an electron microscope, electrons stream off a hot filament and are accelerated to high velocity by passing between two electrically charged plates. The electrons' wavelength is related to their velocity and it is easy to obtain electrons with wavelengths near 0.1 nm, which suggests the possibility of obtaining resolution of about the same magnitude. After the electrons pass through the biological sample, they are focused by magnetic fields in the body of the microscope, those magnetic lenses playing the same role as glass lenses on an optical microscope. Finally, the electrons that were not absorbed by the biological sample hit a luminescent screen and form an image of the sample. The image can be viewed directly or photographed.

The theoretical resolution limit of roughly 0.1 nm usually cannot be attained in practice because of various aberrations similar to those found in optical microscopy. Nevertheless, the change from light microscopy to electron microscopy provides roughly a 1000-fold improvement in resolution.

We could ask, "Why not accelerate the electrons even more and obtain shorter wavelengths, and therefore better resolution?" In fact, electrons with shorter wavelengths ***could*** be obtained. Further, one could obtain a beam of protons, which are much more massive than electrons and which therefore would have short wavelengths even at low velocities. The problem is that either of these high energy beams would quickly deposit enough energy to destroy the object being observed. This problem is one that we encounter again and again at the submicroscopic level — very close observation changes the thing being observed. The electron microscope is thus a compromise between high resolution and target destruction. (Because of the extent to which electron microscopy requires tampering with the specimen, conversations between electron microscopists are peppered with comments like: "You were only looking at artifacts.")

Applications, Further Discussion, and Additional Reading

1. A good source of information on "matter waves" is *Fundamentals of Physics,* 2nd ed. extended, by Halliday, D. and Resnick, R., John Wiley & Sons, New York, 1981, pp. 798–807. A second source is *Introduction To Modern Physics,* 2nd ed., by McGervey, J. D., Academic Press, Orlando, FL, 1983, pp. 106–111.

2. A discussion of the capabilities of various kinds of microscopes, including the electron microscope, is found in "Advances in microscopy," by Root, M., *BioScience,* April 1991, pp. 211–214.

Chapter 8
The Nucleus

Nuclei Are Held Together by the Nuclear Force

A very short-range attraction acts as the "glue" to bind protons and neutrons into a nucleus. This ***nuclear force*** is only effective over distances of about 10^{-15} m, which therefore must represent the approximate diameter of the nucleus. The nuclear force is quite strong — so strong that by comparison the electrostatic repulsion between the protons is negligible. Further, the nuclear force not only binds protons to each other, but also binds protons to neutrons and neutrons to neutrons.

Several Parameters Can be Used to Identify Nuclei

The number of protons in an element's nucleus is the ***atomic number*** and the sum of the protons and neutrons is the ***mass number.*** Nuclei with equal atomic numbers but different mass numbers are ***isotopes*** of one another — they are the same element, but have different numbers of neutrons. We will adopt the convention of appending the mass number to the name or the symbol for an element: carbon has the atomic number 6 and if there are 7 neutrons in the nucleus, the mass number is 13. We will identify this isotope of carbon as carbon-13 or ^{13}C.

Certain naturally occurring isotopes of some elements are able to blacken photographic

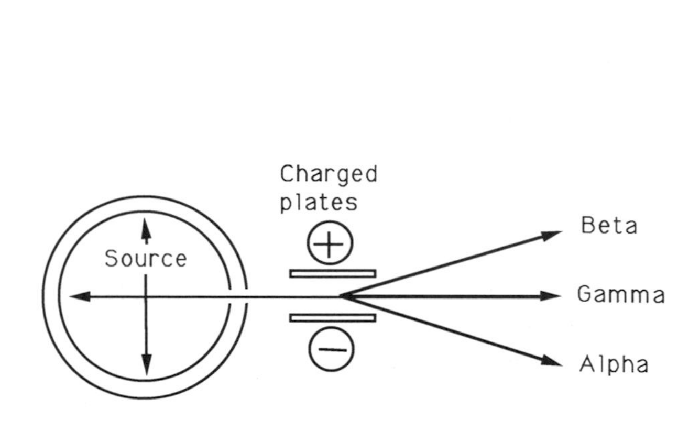

Figure 8.1
Behavior of alpha, beta, and gamma radiations passing through an electric field. Each kind of radiation is deflected according to its charge. In a quantitative description the deflection would also be affected by the particle's kinetic energy.

paper merely by being stored in the dark near the paper. Evidently the isotopes are spontaneously releasing some kind of radiation; they are said to be "radioactive". By passing the emitted radiation between two electrically charged plates, as shown in Figure 8.1, the emissions will often divide themselves into three distinguishable groups: negatively charged, positively charged, and uncharged (although no one isotope would emit all three).

Close study of the particles emitted by a variety of naturally radioactive isotopes ("natural radionuclides") yields this data:

Name	Charge	Rest mass (units of proton mass)	Velocity	Actual identity
Alpha (α)	+2	4	Depends on energy	Helium nucleus
Beta (β)	−1	1/2000	Depends on energy	Electron
Gamma (γ)	0	Never at rest	Speed of light	Electromagnetic radiation

This list is not exhaustive. Other emissions are possible, especially if the radionuclide is artificial, e.g., one created by collisions between naturally occurring nuclei and various sub-atomic particles in a laboratory. Any particular radionuclide decays by emitting one or more identifiable particles at specific energies or in a specific distribution of energies.

There is a further physical parameter specific to a radionuclide: the ***half-life,*** which is the time it takes for half of any quantity of a radionuclide to decay radioactively. The half-lives of various radionuclides span a very wide range, from fractions of a second to billions of years. It is therefore generally possible to identify a particular radioactive isotope from knowledge of its half-life, its emitted particles, and their energies.

An example of a naturally occurring radionuclide is thallium-206 (^{206}Th). It decays spontaneously by the emission of a beta particle (an electron) to yield lead-206 (^{206}Pb), with a half-life of 4.2 min:

$$^{206}\text{Th} \rightarrow {}^{206}\text{Pb} + \text{electron}$$

An example of the creation of an artificial radionuclide is the formation of nitrogen-13 (^{13}N) from a collision between boron-10 (^{10}B) (which is stable) and an alpha particle (a helium nucleus). Note that a neutron is also produced, thus conserving the mass number:

$$^{10}\text{B} + {}^{4}\text{He} \rightarrow {}^{13}\text{N} + \text{neutron}$$

^{13}N is radioactive with a half-life of 10 min, spontaneously emitting a positron (a positively charged electron):

$$^{13}\text{N} \rightarrow {}^{13}\text{C} + \text{positron}$$

Nuclear Isotopic Substitutions and Biological Systems

Few molecular biophysical properties are determined at the nuclear isotopic level, meaning for instance that the chemistry of ^{14}C and that of ^{12}C are essentially the same. There are some interesting exceptions to this statement.

Deuterium (D) is the hydrogen isotope, ^{2}H, and it can be used to make heavy water, symbolized D_2O. An animal whose only source of water is D_2O will soon die. Clearly the isotopic substitution of ^{2}H for ^{1}H makes a very big difference to the animal's metabolism. Water is both an acid and a base and its dissociation is given by:

$$H_2O \rightarrow H^+ + OH^-$$

or

$$D_2O \rightarrow D^+ + OH^-$$

The free hydrogen (or deuterium) ion may then attach to an organic acid:

$$ROOH \rightarrow H^+ + ROO^-$$

followed by

$$ROO^- + D^+ \rightarrow ROOD$$

The question now is, "Is the behavior (in terms of dissociation) the same for ROOD and ROOH?" We should suspect that it is ***not*** because the process of dissociation requires that the proton (or deuteron) moves away from the rest of the compound. A deuteron is twice as massive as a proton and therefore should not move with the same facility as a proton. We can conclude that ROOH and ROOD are not physically the same acid and thus the chemistry of the mouse who consumes D_2O will not be the same as that of a mouse who consumes H_2O. This problem, the decreased mobility of a deuteron compared to that of a proton, can be expected to arise in many other physiological effects whenever D is substituted for H, e.g., the diffusion of deuterated compounds will be slower than that of their protonated relatives. Thus, the biophysical chemistry of an animal fed enough D_2O is lethally changed.

A second nuclear isotopic effect in molecular biophysics occurs when a replicating cell incorporates a radioisotope as part of an important macromolecule. For example, ^{31}P, which is stable, accounts for most of the phosphorus in the world, but there are trace amounts of radioactivc ^{32}P which can be incorporated into deoxyribonucleic acid (DNA). ^{32}P has a half-life of 14.3 days, emitting a beta particle to become sulfur-32 (^{32}S) whose chemical bonding properties are not the same as those of phosphorus. Thus, the decay of ^{32}P in DNA causes that polymer to be broken, leading to genetic consequences.

A third biophysical consequence of nuclear isotopic substitution is that it alters the mutual vibrational behavior of the two nuclei at the ends of a chemical bond. Changes in the energy of bond vibrations are quantized and are caused by the absorption and emission of specific energies of electromagnetic radiation (called infrared, IR; approximately 0.1 eV). The specific IR energy absorbed or emitted by a bond depends on the masses of the two nuclei and on the nature of the bond itself, i.e., covalent or H bond. For example, suppose a cellular process involves the removal of a covalently bonded H atom; the substitution of deuterium at that position will change the vibrational energy of the bond, resulting in a different energy requirement for bond breakage or dissociation. More will be said about this at the end of Chapter 14.

Applications, Further Discussion, and Additional Reading

1. The stability of a nucleus is dependent on the relative numbers of its component neutrons and protons. The ratio of neutrons-to-protons for stable nuclei fits into a very narrow range; outside this range nuclei are unstable and will adjust either the number of protons or the number of neutrons by radioactive decay. In the example given earlier, ^{13}N has 6 neutrons and 7 protons, an unstable combination. Thus, an ^{13}N nucleus emits a positron (charge = +1) to yield ^{13}C, which has 7 neutrons and 6 protons, a relatively stable combination. There is a good discussion of this concept in *University Physics*, 5th ed., by Sears, F. W., Zemansky, M. W., and Young, H. D., Addison-Wesley, Reading, MA, 1980, pp. 791–801.
2. The half-lives of radioactive nuclei range from fractions of a second to many millions of years. Most radioactive nuclei created during the evolution of the solar system have had plenty of time to decay to harmless elements, but quite a bit is still around, e.g., in the form of uranium, radium, and radon. The rock aggregate used to make concrete building blocks in the southeastern U.S. often has an abnormally high radioactivity.
3. The ability of an emitted particle to penetrate into a substance is determined by a quantity called the ***linear energy transfer*** (LET). LET is a measure of how much energy a particle gives up per unit of track length in units of, say, electron volts per 10 nm. A high LET particle deposits

energy densely in its wake and will therefore have only a short track length, which is synonymous with low penetration distance. LET is highest (penetration is least) for particles with low velocity and high charge. As a result, an alpha particle has a much higher LET than does a beta particle of the same energy. In fact, most alpha particles cannot penetrate the dead layer of our skin and are therefore harmless if administered externally. On the other hand, an inhaled or ingested alpha emitter can make direct contact with the living cells of the lung alveoli or of the gastric and intestinal epithelia.

Uranium-238 (^{238}U) decays radioactively into another radionuclide which also decays; after several such steps radon-226 (^{226}Rn) is formed. Thus, ^{226}Rn is called a "daughter" of ^{238}U. Radon is unusual in that it is a natural radionuclide and is also a gas, and when released from geological formations by seismic activity, mining, or home building it can be inhaled or ingested. If nothing further happens, the radon will be exhaled, excreted, or defecated because it is in the same column of the Periodic Table as helium and argon and is therefore chemically unreactive. The biological problem with radon arises when it is inhaled or ingested and then subsequently decays radioactively in the lungs or intestines of an animal. Among the subsequent daughters of radon in the decay series are two isotopes of polonium, which emit energetic alpha particles. Polonium is in the same column of the Period Table as oxygen and sulfur and can react with tissue components, thus localizing its effect to small regions of the lungs and intestines. The alpha particles emitted by tissue-fixed polonium atoms are primarily responsible for the high lung-cancer rates that were found in uranium miners before safety precautions concerning radon inhalation became common in that industry. We see that radon's fearsome reputation is actually due to radon ***daughters,*** the most immediate problem with the radon itself being its gaseous nature.

4. Nuclear reactors generate large quantities of concentrated radioactive wastes. This is especially true of "breeder" reactors, which artificially produce more radioactive material than goes into them in the first place. This waste presents a serious biological hazard, the reduction of which is usually in the form of isolation and storage because there is no reliable way to speed up the decay process. (Artificial transmutation to a less dangerous nuclide is prohibitively inefficient.)

   Consider that the storage of radionuclides will have to be maintained for many half-lives, perhaps millions of years, that radioactive wastes are very corrosive, and that there is no way to pretest a storage system for so long a period of time. Thus, we must rely on unproven storage technology, an act which verges on the religious, among other things.

   The only way to test a potential "million-year" disposal system is to assume ***reciprocity.*** This means that the system is exposed to a million units of radiation for 1 year, say, and then the effect on the system is assumed to be the same as if the system were exposed to one unit of radiation for a million years. In fact, there are many examples of radiation effects which do ***not*** exhibit reciprocity or in which reciprocity has not been demonstrated. (The assumption of reciprocity is a common one, lying at the heart of much testing of, and therefore conclusions regarding, the long-term effects of pollutants, drugs, carcinogens, etc.) You can find a good discussion of radioactive waste disposal and of the consequences of improper disposal in *Nuclear Waste: The Problem That Won't Go Away,* Worldwatch Paper, Number 106, The Worldwatch Institute, 1776 Massachusetts Avenue, N.W., Washington, D.C., 20036.

5. There are numerous practical ***applications*** of isotopic differences in the study of living systems. For example, ^{14}C, a radioactive isotope produced in the upper atmosphere by cosmic radiation, is incorporated into living systems along with the stable isotope ^{12}C. At death the organism stops incorporating both isotopes; thus, the ^{12}C level subsequently remains the same while the ^{14}C decreases because the latter is radioactive and is no longer being created in the organism's body by cosmic radiation. Thus, the elapsed time since the organism died can be measured by the remaining proportion of the two isotopes (carbon dating).

   Second, radionuclides which localize at the site of a tumor can be used to help destroy the tumor. For example, iodine localizes in the thyroid gland (it is a component of hormones produced there); thus, radioactive iodine can be used to treat thyroid cancer.

Third, the energy, charge, and mass of the particle emitted by a radioisotope can be used to identify the isotope itself and its location. In the famous Hershey and Chase experiment, radioactive sulfur was used to label a bacteriophage's protein specifically (DNA does not contain sulfur) and radioactive phosphorus was used to label the bacteriophage's DNA specifically (proteins do not contain phosphorus). Hershey and Chase showed that, upon replication, the inheritance of phage genetic properties followed the phosphorus, not the sulfur, thus establishing the role of DNA rather than protein as the carrier of genetic information. Similar applications are discussed in *Intermediate Physics for Medicine and Biology,* 2nd ed., by Hobbie, R. K., John Wiley & Sons, New York, 1988, pp. 475–510.

6. The biological effects of deuteration are discussed in *Biological Effects of Deuterium,* by Thomson, J. F., Macmillan, New York, 1963.

7. There are extensive nonmathematical discussions of nuclear physics, as it applies to radioactivity, biological effects, cosmology, nuclear energy, and nuclear war in *Radiation and Radioactivity on Earth and Beyond,* by Draganic, I. G., Draganic, Z. D., and Adloff, J.-P., CRC Press, Boca Raton, FL, 1989.

8. You can find a comprehensive discussion of the effects of radiation on living systems in *Biological Effects of Radiation,* by Coggle, J. E., International Publications Service, Taylor and Francis, New York, 1983.

Chapter 9
The Atom — The Plum Pudding Model

An Outmoded-But-Useful Model

At the beginning of this century, J. J. Thomson, the discoverer of the electron, suggested a model of the atom in which the electrons were small individual negative charges imbedded in a diffuse positive charge; in other words, there was no nucleus as such. Rather, the positive charge we now associate with the nucleus was supposed to be a kind of cloud. This "plum pudding" model, where electrons were analogous to raisins in a fluffy positively charged cake, explained the electrical neutrality of the atom — the sum of the electrons' negative charges matched the total positive charge (see Figure 9.1).

The plum pudding model suggested an easily interpretable experiment: a beam of alpha particles from an alpha emitter should pass right through the plum pudding with little bending of the beam. The reasoning was that alpha particles were known to have a charge of +2 and a mass of 4. An alpha particle incident on a plum pudding-type of atom would experience little electrical repulsion from the atom's diffuse positive cloud and would easily bump aside the much less massive electrons. Thus,

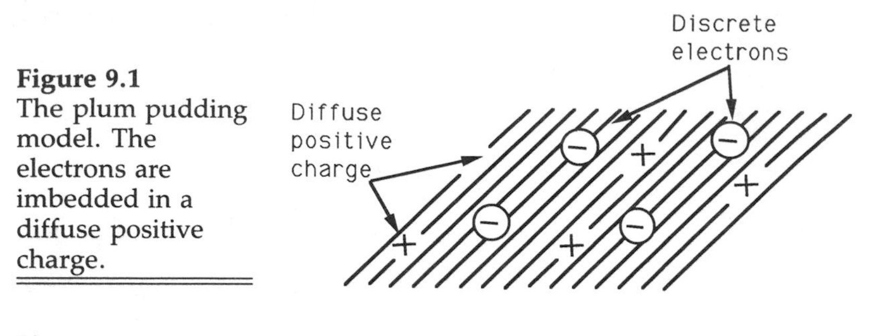

Figure 9.1
The plum pudding model. The electrons are imbedded in a diffuse positive charge.

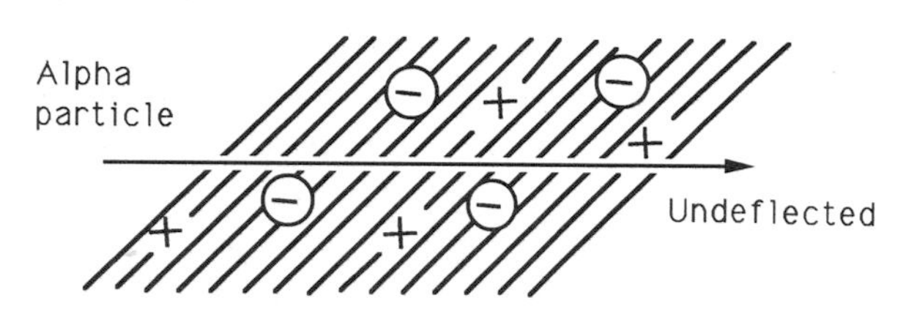

Figure 9.2
Behavior of an alpha particle penetrating a plum pudding atom. The massive alpha particle will hardly be deflected because it will bump the light electrons aside and go right through the diffuse positive charge.

the alpha particle should experience little deflection while traversing a group of plum pudding-type atoms. This is shown in Figure 9.2.

The experiment was run by Rutherford. As a target he used gold foil, which can be pounded into very thin sheets, a feature necessitated by the fact that alpha particles barely penetrate matter. After the alphas hit the gold their angle of deflection was measured by observing the point at which they impacted a scintillation screen. The experiment is diagrammed in Figure 9.3.

The plum pudding model predicted small deflections, thus most scintillations on the screen should be near point A. In fact, Rutherford found many collisions near point B, a result completely inconsistent with the plum pudding model. Collisions at B are due to 180-degree

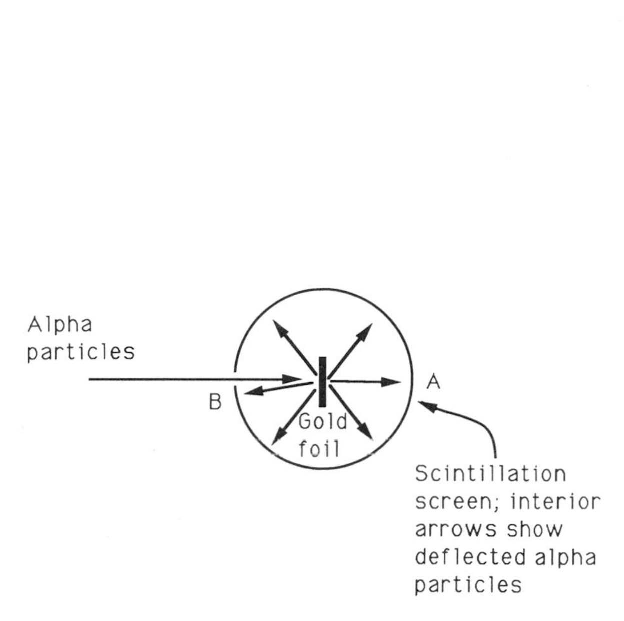

Figure 9.3
The equipment for Rutherford's experiment. An observer sees scintillations where alpha particles hit the screen. Undeflected particles hit the screen at A, but many particles were deflected backward to B, indicating that they had encountered a concentrated positive charge in the gold foil.

deflections of the alpha particles and could only result from an encounter between an incident alpha particle and a very heavy, condensed positive charge in the gold atoms. This tiny cluster of positive charge is what we now call the nucleus of the atom.

The Thomson model was presented here to show that it was wrong but useful, in the sense that it suggested an experiment leading to its own demise and to a better model. The Thomson model is only one in a string of models for the atom, each in turn giving way to a better description. We should always be very careful about assuming that today's idea is the final word; science never has been like that — just as with politics and hemlines.

Applications, Further Discussion, and Additional Reading

1. There is a somewhat technical presentation of the Thomson model in *The Structure of Matter: A Survey of Modern Physics*, by Gasiorowicz, S., Addison-Wesley, Reading, MA, 1979, pp. 132–146.

2. The inability of alpha particles to penetrate very far into matter, such as the gold foil in Rutherford's experiment, is due to the high rate at which they deposit energy in their wake. They are said to have a high ***linear energy transfer,*** or LET, which is a measure of energy deposited per unit length of track. In fact, the LET of alpha particles is so high that they cannot penetrate the dead layer of cells, the stratum corneum, of our skin; their energy is dissipated too quickly. Thus, we are protected from alpha-particle emitters external to us. On the other hand, if we breathe or ingest alpha-particle emitters, that places the emitters right next to the cells of our lungs and digestive tracts, which have no stratum corneum. In calculating absorbed doses of radiation, health physicists therefore note carefully the amounts of alpha emitters in the air and food.

Chapter 10
The Atom — The Bohr Planetary Model

The response to Rutherford's work was the so-called planetary model of the atom — electrons revolving around the massive small nucleus like planets around the sun. The implications of this model are manifold and require explanation.

Tangential and Radial Motions Keep a Satellite in Orbit

First, we need to understand why a satellite or planet stays in orbit. The satellite in Figure 10.1 potentially is capable of motion in two separate directions. If there were no pull by gravity toward the earth, then the satellite would fly off into space tangentially to the orbit; this motion is called the ***tangential*** velocity. On the other hand, if the tangential velocity were zero, the satellite would move straight toward the earth because of gravitational attraction; we say the earth exerts a ***radial*** force. Thus, the satellite's motion is due to the constant "correction" of the tangential velocity by the radial force. Figure 10.2 shows these two effects as occurring in small separate acts (although, of course, the actual motion is smooth). In every small time period, the satellite can be imagined as moving tangentially and radially, resulting in smooth orbital motion.

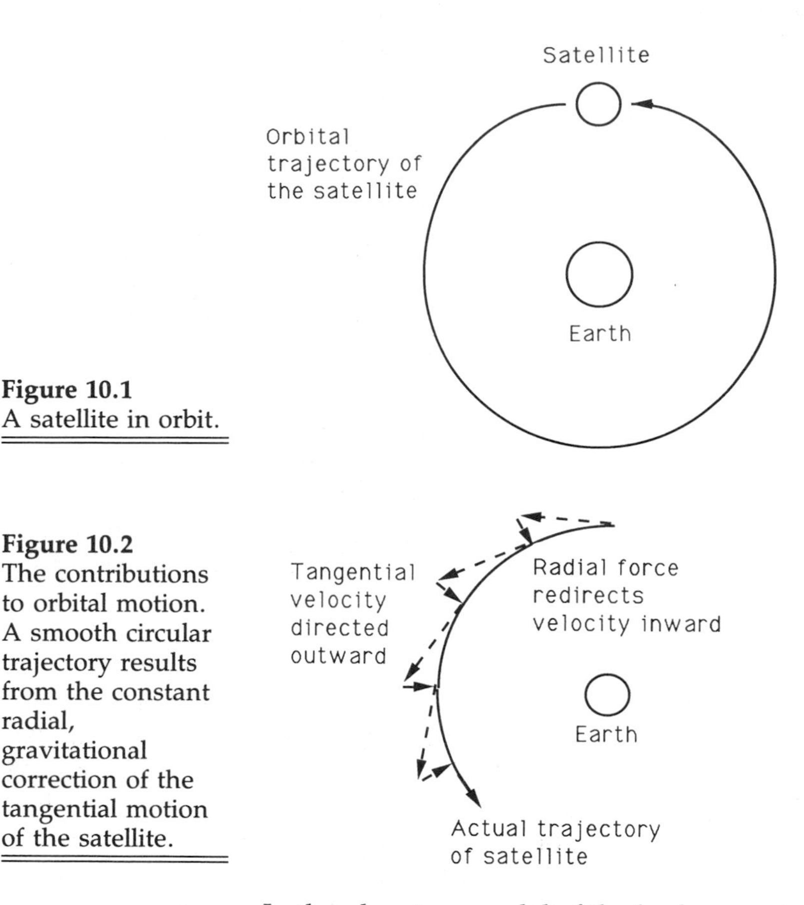

Figure 10.1
A satellite in orbit.

Figure 10.2
The contributions to orbital motion. A smooth circular trajectory results from the constant radial, gravitational correction of the tangential motion of the satellite.

In the planetary model of the hydrogen atom, the electron orbits the nucleus (a proton). The radial force is thus due to electrostatic attraction rather than to gravity, but otherwise the description is the same as for the planetary system of Figure 10.2 — a satellite of low mass is held in orbit around a massive central object by the dual effects of tangential velocity and radial force.

The Planetary Model Answered Some Questions and Raised Others

The planetary model clearly yields zero net charge for the atom and, by having a well-defined nucleus, also explains Rutherford's data on alpha particle scattering. It also opened a can of conceptual worms that could not be dealt with by the physics of the time.

The 19th-century physicist Maxwell had derived a set of equations which described the behavior of, and relationships between, electric and magnetic phenomena. There were extensive experimental data to back up the predictions of Maxwell's equations. One of the predictions of these equations was described in Chapter 3: an electric charge, when accelerated, should radiate energy. In the planetary model an orbital electron has an electrostatic force attracting it to the nucleus and Newton's Second Law associates an acceleration with every force. Therefore, an electron in orbit is being accelerated and should radiate energy, causing the orbit to get smaller and smaller as more and more energy is radiated away from the electron. Finally, the electron should crash into the nucleus and the whole planetary structure would be lost. This "spiral catastrophe" seemed to make a planetary atom untenable.

Bohr stepped in with a revolutionary proposal, namely, that electrons ***did*** have stable orbits, Maxwell's equations notwithstanding. Bohr then went even further by suggesting that the radii of those stable orbits were restricted to certain specific values. Both parts of this proposal seem to defy our common sense: the first part, stable orbits, is inconsistent with the well-tested Maxwell equations, which predict the spiral catastrophe. The second part, restricting orbital radii to specific values, flies in the face of our experience with artificial earth satellites. The height of the orbit of a satellite above the earth can be adjusted and fine tuned to any degree merely by judicious firing of small thruster rockets on the satellite. In principle any orbital radius whatsoever can be obtained without restriction. Bohr's hypothesis suggested that this reasoning

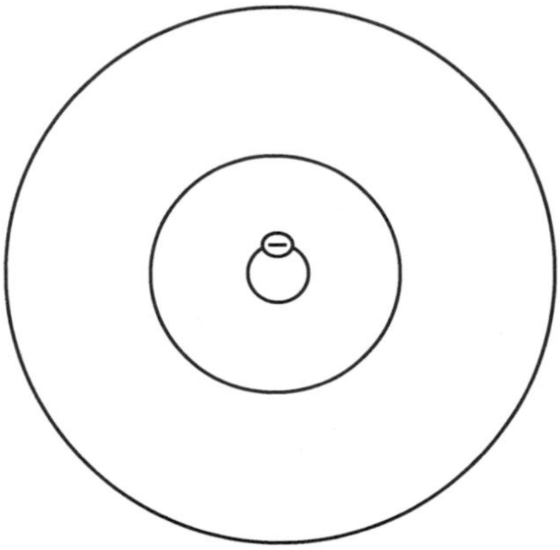

Figure 10.3
The hydrogen atom in its ground state; the first three Bohr orbits are shown to scale; the nucleus is omitted. The electron is indicated by the encircled minus sign.

just wasn't applicable to electrons, although he presented no *a priori* theoretical grounds for believing this.

A diagram of a Bohr-type planetary model of the hydrogen atom is shown in Figure 10.3. Note that several possible orbits are shown, only one of which contains an electron (because the hydrogen atom has only one electron). When the electron is in the innermost orbit of the atom, which is the orbit most tightly bound to the nucleus, the atom is said to be in its ***ground state.***

If sufficient energy were given to the electron it would move into one of the normally unoccupied orbits, in which case we would say that the atom is in an ***excited state.*** This is shown in Figure 10.4.

Energy-Level Diagrams: A Convenient Format

The radii and the energies of electrons in a Bohr atom can be presented in a somewhat different format from the two preceding figures. This arrangement, shown on the right side of Figure 10.5, is called an ***energy-level diagram***. The correspondence between energy levels and orbital radii is shown by the arrows. There are several important points to be made about this energy-level diagram. First,

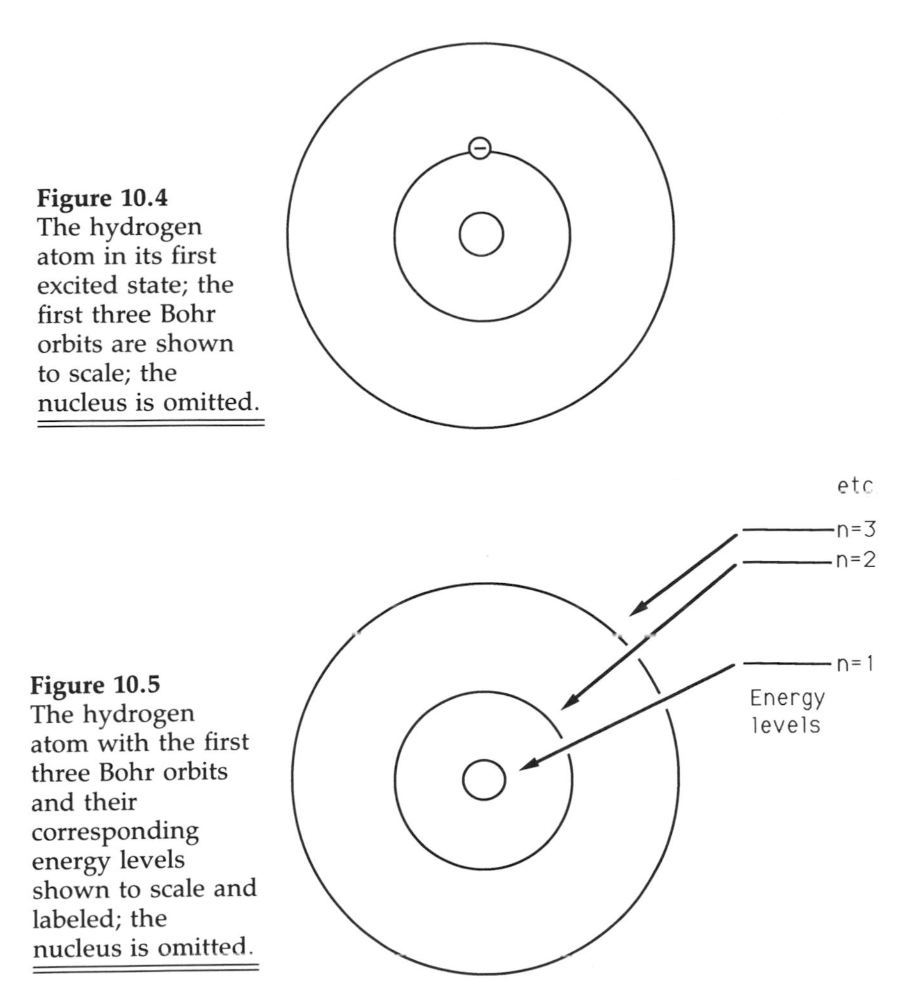

Figure 10.4
The hydrogen atom in its first excited state; the first three Bohr orbits are shown to scale; the nucleus is omitted.

Figure 10.5
The hydrogen atom with the first three Bohr orbits and their corresponding energy levels shown to scale and labeled; the nucleus is omitted.

every level, or orbit, is labeled with a positive integer n, called a ***quantum number.*** Only those orbits with an integral quantum number are allowed by the Bohr model. Second, there is a radius corresponding to every value of n. Thus, no other radii are allowed. Note that the radii get bigger and bigger with increasing n. Third, there is an energy corresponding to every integral value of n. Thus, no other energies are allowed. Note that the energies get closer and closer to one another, finally converging to the value zero. The fact that the

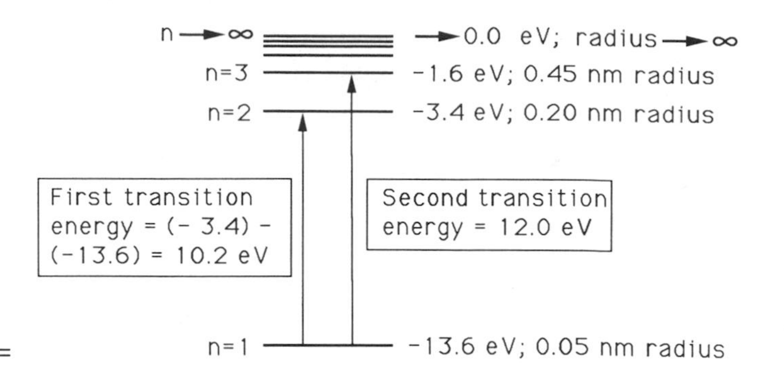

Figure 10.6
A Bohr energy level diagram for the hydrogen atom, showing how the energies converge to 0 eV and radii diverge to infinity.

energies are negative should not distract you — it is only the ***differences*** in energy levels that are important.

Using the above energy-level diagram we can now see that the Bohr theory predicts the correct experimental results for the hydrogen atom. In the ground state the electron will be in level n = 1, having a radius of 0.05 nm and an energy of −13.6 eV. If the electron is given a quantum of electromagnetic energy of 10.2 eV, it should jump to the n = 2 level, having a radius of 0.2 nm and an energy of −3.4 eV. Other transitions to excited states are possible: n = 1 to n = 3 should require the absorption of a 12 eV quantum. Figure 10.6 shows these predicted transitions diagrammatically. In fact, these ***are*** the energies of the light photons actually absorbed by hydrogen atoms in experiments, constituting a powerful confirmation of the Bohr model.

We might now ask what would happen if we irradiated hydrogen atoms with, say, 11.0-eV light. The answer is, "absolutely nothing"; the light would sail right through the hydrogen without any effect because there is no energy level 11.0 eV above the n = 1 level. Such transitions are said to be ***forbidden***. Especially important is the fact that there would be no partitioning of the light energy into 10.2 eV

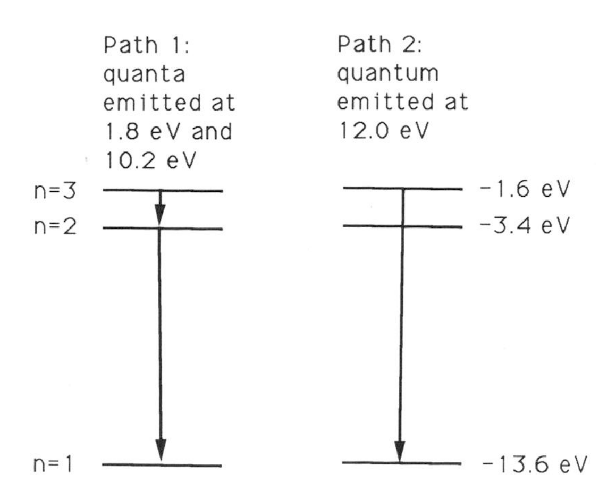

Figure 10.7 The Bohr hydrogen atom, showing light emission via two different paths following excitation to the n = 3 level. The 1.8-eV transition in path 1 is not seen in absorption, but was predicted by the Bohr model for emission.

for an n = 1 to n = 2 transition, with a 0.8 eV quantum left over; the quantum can only be absorbed as a unit. It has been experimentally verified that only those transitions between levels with positive integral n are allowed.

This then is our picture of light absorption in the Bohr model: a quantum can be absorbed if it has exactly the energy difference between an allowed occupied level and an allowed unoccupied level; the electron will then move to the new, bigger orbit. No other quanta can be absorbed.

Light Emission Is Accurately Predicted by the Bohr Model

The Bohr model also predicts light emission. If excitation is caused from the n = 1 to the n = 3 level, the electron will subsequently return to the ground state by one of two paths, shown in Figure 10.7, emitting light quanta at each step. The transition from n = 3 to n = 2, for example, causes the emission of a quantum whose energy is not associated with any absorption process because the n = 2 level is not occupied in the ground state hydrogen atom.

Figure 10.8
The Bohr model of a polyelectronic atom. Because of the large positive charge of the nucleus, the energy level separation for the innermost orbits should be much greater than for hydrogen. Energies of up to 250 keV are common, compared to about 10 eV for hydrogen.

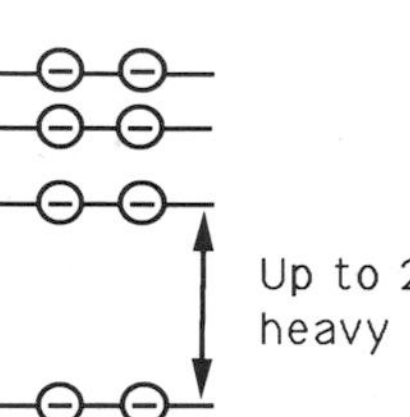

The Bohr Model Qualitatively Predicts X-Rays

It is difficult to extend the Bohr model to atoms having more than one electron. The reason is that electron-electron repulsions appear in polyelectronic atoms, and those interactions cannot be built into the Bohr model in a reasonable way. However, the model can still be used *qualitatively* to explain the generation of X-rays from polyelectronic, or heavy, atoms. The simplest way to model a heavy atom in the Bohr model is by filling up the Bohr levels, as shown in Figure 10.8.

The innermost electrons of a heavy atom are held to the nucleus much more tightly than the electron of hydrogen because the polyelectronic atom's nucleus has a very large positive charge. As a result, the energy difference between the $n = 1$ and $n = 2$ levels may be several hundred keV (keV = 1000 eV). We can knock out an $n = 1$ electron by bombarding the heavy atom with a high-energy electron beam, as shown in Figure 10.9. This leaves a vacancy in the $n = 1$ orbit, into which the $n = 2$ electron can fall, causing the emission of a quantum of electromagnetic energy of several hundred keV. This radiation is called X-rays. A medical X-ray machine generates radiation in this fashion.

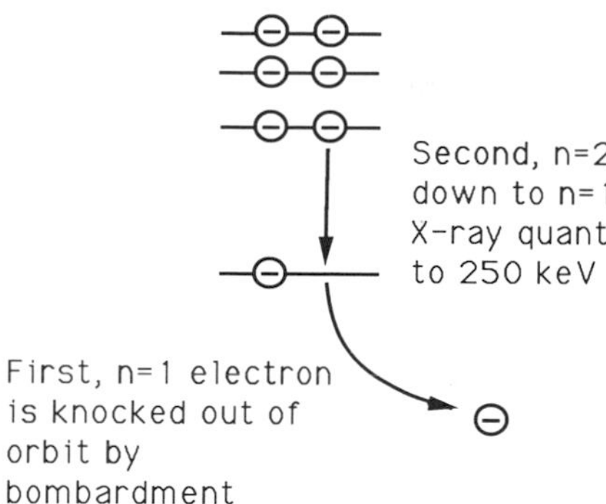

Figure 10.9
The Bohr model of a polyelectronic atom, showing the origin of X-rays from heavy atoms.

IONIZATION ENERGIES ARE PREDICTED BY THE BOHR MODEL

The Bohr model has one more surprise for us. Refer back to Figure 10.6 and note that for large values of n the orbital radius becomes infinite. We can interpret the idea of an "infinite" radius as meaning that the electron and proton are so far apart that they no longer interact electrically in any way. We then say that the hydrogen atom has been ***ionized.*** However, examination of Figure 10.6 shows that movement to "infinite radius" requires only a finite energy — absorption of a 13.6 eV quantum. In fact, that is experimentally what it takes to ionize the hydrogen atom — another confirmation of the Bohr theory.

Ionization provides us with one consideration not associated with excitation: absorbed energy ***in excess of the ionization energy*** can be partitioned into electron kinetic energy. Absorption of a 13.6 eV quantum by hydrogen moves the electron to infinity (read, "far away"), at which point the electron has zero energy, all of the absorbed energy having been used up in the ionization process. By "zero energy" we mean that the electron has no kinetic energy because it is not moving and that it has no potential energy because the nucleus is too far away to interact with it.

Suppose now that a 14.6-eV quantum is incident on the hydrogen: ionization will still occur, removing the electron to infinity but leaving the electron with the excess of 1.0 eV (14.6 − 13.6 eV) in the form of kinetic energy. In other words, the electron will move about slowly, but will be far removed from the parent proton. Alternatively, absorption of a 30-eV quantum causes ionization and yields an electron with 16.4 eV (30.0 − 13.6 eV) of kinetic energy. What these numbers mean is the absorbed energy is partitioned between the ionization process and kinetic energy, but the quantum is nevertheless absorbed as a unit. Thus, the quantum picture of light is still quite valid for ionization.

There Are Problems with the Bohr Model

The Bohr model predicts the absorption and ionization energies of the hydrogen atom, thus apparently vindicating Bohr's radical assumptions of stable, specific orbits. That, however, is about as far as we can push this model because when we try to apply it to helium (atomic number = 2) several problems appear. One of these problems is obvious in any case, one is obvious only on first reading, and one, not obvious at all, strikes at the heart of our everyday experience.

The first problem is easy to see: the two electrons of helium must electrostatically repel one another, a factor not relevant to the case of hydrogen. This factor plays a key role in determining the ionization and excitation energies of helium and cannot be accurately built into the Bohr model. This, in turn, suggests that interactions between electrons are not as simple to describe as "two negative point charges pushing at one another".

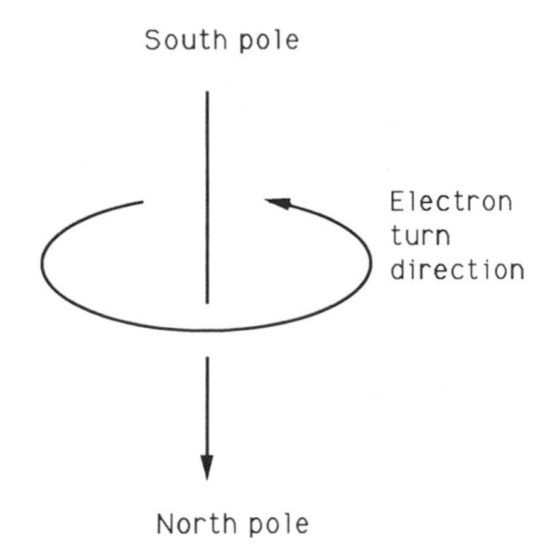

Figure 10.10
A magnet created by electron spin.

The second problem only seems easy to see: electrons appear to have an intrinsic spin. In the planetary model we can regard this spin as the analog of the earth's 24-hr rotation. The problem with this simple picture is that spin can be ignored in the interaction of planets, but can't be ignored for electrons' interactions.

Figure 10.10 shows how the electron spin is detected. It is well known from classical physics that a moving electric charge generates magnetism. The spinning electron can be thought of as an electric charge moving in a circle: its consequent magnetism can be detected by appropriate instruments.

The two small magnets created by a pair of electrons will interact as any two magnets would, but ***that*** effect actually is small. A real problem — known from deeper considerations — is that no two interacting electrons can have exactly the same physical state. Thus, if both of the helium electrons were in the n = 1 energy level, they would be forced to have different spin directions in order to have different overall states. Bohr's model does not account for the spins of electrons; however, we know that they must be taken into account in order to predict accurately excitation and ionization energies of polyelectronic atoms. This topic will be discussed in more detail in Chapter 11.

The third problem really defies our senses: how does an electron travel from one allowed energy level to another if, according to Bohr, the intermediate radii are forbidden? Suppose an electron is in the n = 1 level and absorbs a quantum with exactly the correct energy to move to the n = 2 level. We will know if a transition occurs because the quantum will disappear or because the excited state will reveal its presence; for example, the excited atom can return to the ground state by emitting a quantum of the same energy, or the excited state may react chemically in a manner different from an electron in the ground state, or the excited state may decay by releasing heat energy. Thus, we can be certain that the n = 1 to n = 2 transition has occurred and that the electron has moved between the corresponding radii. Now we ask, "What path does the electron take in moving between the two states?" We cannot assume that the electron merely jumps directly from the one orbit to the other because it would then have to pass through radii forbidden by the Bohr theory. We are therefore forced to take a viewpoint that the electron disappears from one orbit and then reappears in the other. Nothing in our experience prepares us for that kind of conclusion, but that's the way it is! The three considerations just discussed lead us to another model of the atom in the next chapter.

Applications, Further Discussion, and Additional Reading

1. Coverage of the topics of orbital motion, the Bohr atom, and X-ray generation can be found in *University Physics,* 5th ed., by Sears, F. W., Zemansky, M. W., and Young, H. D., Addison-Wesley, Reading, MA, 1980, pp. 89–106 and 753–764.

2. An application to radiation biology: We can now understand why the absorption of a 250-keV X-ray or a 1-meV γ-ray can have such serious biological consequences. Absorption of the γ-ray by a biochemical, say, causes ionization and liberates an electron with almost 1 meV of kinetic energy — a very energetic electron indeed. This fast electron then speeds through a cell, causing enormous damage through subsequent ionizations and excitations along its track.

   It can be shown that an electron has quantized energy levels ***only*** when it is bound, or confined, e.g., in an orbit. Therefore, absorption of energy from a passing, or free, incident electron is not restricted to the entire energy of the electron — energy can be given up in any amount possessed by the incident electron. An incident electron thus is capable of causing ionization or excitation of any atom (as long as the atom's lower electronic level is occupied and the upper one is not). Ionizations and excitations thus created lead to new chemical species that lack the correct biological functions or have harmful biological consequences, such as cancer or deleterious mutations.

Chapter 11
THE ATOM — THE QUANTUM MECHANICAL MODEL

A THOUGHT EXPERIMENT TO DEMONSTRATE THE UNCERTAINTY PRINCIPLE

In Chapter 1 we encountered the notion that the observer really is an intrinsic part of the observation process and that this is especially important when observing submicroscopic phenomena. The process of observation disturbs the thing being observed and that disturbance might possibly affect other observations. The example was made of using a scintillation screen to detect the position of an electron; once it hit the screen we would never ***again*** be able to observe the electron.

We now consider a "thought experiment" to observe an electron in orbit in the n = 1 state. We shine light on the electron to "see" it and, of course, must use a minimum of one photon. However, the least energy to which the electron will respond is of the size necessary to cause the transition from n = 1 to n = 2; any other photon will sail right past the electron without any change in the light or the electron. Thus, in order to "see" the electron we must excite it to the n = 2 level, which means it will no longer be in the original orbit we wanted to observe. The upshot is that we will have been able to identify at most ***one*** point in the n = 1 orbit before losing track of that orbit altogether. In order to plot any trajectory we require at least ***two*** points that we can connect by a line, and we see now that getting a second point was precluded by our getting the first point.

Two ideas are noteworthy here. ***First,*** in order to observe the electron we had to use a photon whose energy was at least comparable to that of the electron, thus disturbing the electron profoundly. We could observe an automobile using a photon of the same energy as before and it would not have any perceptible effect on the car at all. Evidently the effect of the observation process normally becomes small at the macroscopic energy scale. ***Second,*** the problem of disturbing the electron cannot be solved by building different equipment because the least energy to which the electron will respond is inherent in the structure of the atom itself and has nothing to do with the equipment.

Some Measurements Can Interfere With Each Other

We first need to examine the word "uncertainty"; it means the extent of our ignorance of the "correct" value and it results from the observation process, not from engineering shortcomings. Note the reference to ***our*** ignorance — it implies that the uncertainty is intrinsic to the observer.

Heisenberg showed that ***simultaneous*** measurements of an electron's position and its momentum interfere with each other and thus induce essential uncertainties into those measurements. In order to draw an electron's trajectory we must know where the electron ***is*** (position) and, simultaneously, where it is ***headed*** (momentum); the Uncertainty Principle forbids us from having that knowledge.

Thus, we cannot assign a trajectory to the electron and must discard the planetary model. In keeping with the analogy of the previous paragraph, we must approach this as though the electron does not have a well-defined position and momentum simultaneously. In any case, there is no way to measure both of those parameters simultaneously and accurately.

We have now accumulated several problems with the planetary electron model. They are the spiral catastrophe, the question of how the electron traverses the forbidden space between orbits during a transition, the specific consideration of the thought experiment on observing the electron, and general considerations of the Uncertainty Principle. At the heart of these problems is Bohr's radical, ***ad hoc*** assumption of stable, discrete orbits — an assumption essentially justified by the fact that it worked in predicting spectroscopic properties of hydrogen. It would be intellectually more satisfying if the idea of stable, discrete orbits had arisen naturally from a theory with physically simpler assumptions, but there was no such theory. Instead, a new, even more radical theory was later advanced, in which the concept of discrete orbits was completely abandoned.

Probability Clouds Replace Well-Defined Orbits

Quantum mechanics represents the behavior of electrons around nuclei in terms of mathematical constructs called ***wavefunctions,*** or ***orbitals***, which contain all possible information about the electron. In particular, the square of the wavefunction (orbital) gives the probability of finding the electron in a given region of space. What is meant by "probability" is that repeated observations of the electron's position give a ***distribution*** of positions in space, rather than a single point or a

Figure 11.1
An electron probability cloud; the smooth contour encloses the region of space in which the electron is most likely to be found.

well-defined trajectory. The electron is generally found in a region within 0.1 nm or so of the nucleus, but almost never at exactly the same point twice. This peculiar result yields what is called a "probability cloud", or "electron cloud", an example of which is seen in Figure 11.1.

The probability cloud of Figure 11.1 requires some interpretation. It is shown with a sharp outer boundary when, in fact, a "real" probability cloud would extend sparsely beyond that boundary to a long distance. To draw an analogy, imagine a rain cloud whose edges are fuzzy; now draw a smooth spherical surface that encloses, say, 90% of the water vapor in the cloud. You would get a picture somewhat like that of Figure 11.1. The way then to interpret Figure 11.1 is to say that the smooth boundary is a contour that encloses most of the probability of finding the electron, i.e., the electron will most often be found somewhere inside the indicated spherical region, which is centered at the nucleus.

Do not suppose that the electron itself is "smeared out". Rather, think of the enclosed region as the region you would search first if you wanted to find the electron; with repeated observations that is where it would most often turn out to be. We will see shortly that many probability clouds are ***not*** spherical, but in each case a sharp boundary will be shown enclosing most of the probability and defining the shape of the cloud. The region ***outside*** the cloud contains so little probability of finding the electron that we can ignore it.

We note in passing that the notion of a probability cloud really implies a kind of cloudiness in an observer's ability to find the electron. After all, the electron really is ***somewhere***, but the observer will have to search for it, starting with the densest region of the probability cloud. Further, knowledge of where the electron was in a previous observation is of little help in finding it in a subsequent observation, as would be the case if there were well-defined orbits.

The concept of probability clouds frees us from the problems we noted earlier with the planetary model. The first is the spiral catastrophe. There are no distinct orbital trajectories associated with probability clouds, so the electron is not accelerated and no consequent radiation is emitted. Second, without discrete orbits we do not have to worry about how the electron spans the forbidden radii between two orbits following light absorption because there ***are no*** orbits. "Undergoing an excitation transition" merely means that the electron becomes more likely in new regions of space and less likely in the old region. Another way to view the transition is as the fading out of one probability cloud and the fading in of another probability cloud, just like a scene fades in a movie. This does not mean that the electron ***itself*** fades out, but rather that the ***probability*** of finding it in a given region decreases while the probability of finding it in another region increases. Third, the allied considerations of the observer's role and of the uncertainty principle do not contradict the quantum theory.

In fact, the quantum theory builds in both of those considerations in a natural way.

Applications, Further Discussion, and Additional Reading

1. The idea of associating uncertainty with the observer rather than with the thing being observed is a peculiar one. In fact, in ***practical*** terms the two situations are the same, as we can see from a macroscopic analog: suppose we have a bank account of whose balance we are uncertain by $100 because of our arithmetic errors. The uncertainty is in our mind only because there really *is* an actual, exact balance. On the other hand, suppose the uncertainty were due to someone else randomly changing our balance by $100. Our check-writing behavior can then be no different in the two cases, even though the uncertainty originates from two quite different sources.

2. For mathematical reasons, physicists and chemists almost always use the abstract concept of wavefunctions (orbitals), rather than that of electron probability clouds, to describe electrons. We will almost always do the opposite — in keeping with our extensive use of real world models. You need only to remember that squaring a wavefunction yields the probability cloud, so the two are really very closely related.

   Perhaps the association of probability clouds with real world models bothers you. Consider that we run into the notion of probability in many of our life experiences, from weather forecasting to crap shooting. Most people feel comfortable with statements like "There's a 20% chance of rain," "I have a good chance to get that date," and "Throwing two ones at dice is unlikely." The same concept applies to the statement "There's a 40% chance that the electron will be found in a particular region of space." It is quite another matter to talk about the ***square root*** of the probability of finding the electron in a particular region of space. A wavefunction yields the square root of a probability, which has no real world significance until it is squared. (In fact, the square root of a probability can be an imaginary number!) Because of this lack of real world significance wavefunctions were called "mathematical constructs" earlier in the text; they are merely intellectual constructions which can be used to describe physical phenomena. In short, they are tools.

3. Observing an electron with a photon whose energy is close to that of the electron generates a considerable disturbance to the electron. Suppose that you observed a moving car with a "particle" having the same energy as the car; what would you use?

4. The distribution of probability ***within*** a probability cloud is not uniform. For some, the greatest probability is near, or even at, the nucleus. For others, the probability is zero at the nucleus.

Chapter 12
THE HYDROGEN ATOM

PREDICTIONS OF QUANTUM THEORY

As discussed in the previous chapter, quantum mechanics describes the electron as a probability cloud. This probability is not uniform, but varies considerably from place to place. It is possible to calculate the region of space in which the electron has the highest probability of being found. For the hydrogen atom, this ***most probable*** radius of the first quantum mechanical electron cloud is identical to the radius of the first Bohr orbit (0.05 nm). This is a very pleasing result; a serious difference between the two radii would have been difficult to explain. In addition, the electronic transition energies from the electron clouds are the same as predicted by the Bohr model. The quantum mechanical description goes much further, however. Not only does it predict a series of increasingly larger spherical electron clouds, but other electron cloud series as well. Because of their three-dimensional shapes, these electron clouds sometimes are said to describe ***space states.*** Examples of two which interest us, the "s" series and the "p" series, are shown below in Figure 12.1.

To put the sizes of these clouds into perspective, their largest dimension is of the order of

Figure 12.1
Two common types of electron probability clouds. The nucleus, too small to be seen at this scale, is at the geometric center of each.

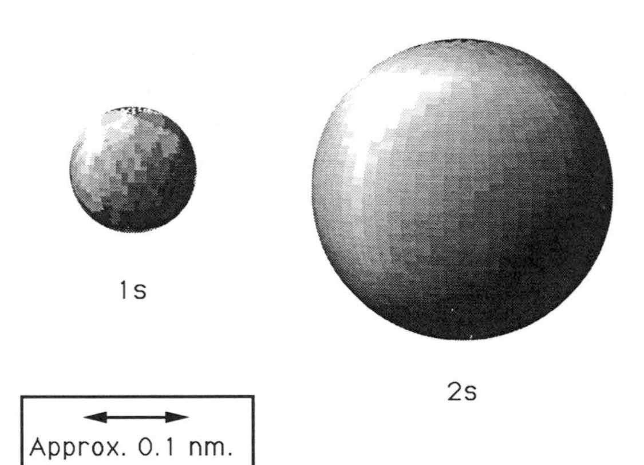

Figure 12.2
The first and second s-type electron probability clouds of hydrogen.

10^{-10} m (0.1 nm), with a nucleus of diameter 10^{-15} m at the center. The word "center" here refers to the geometric center of an s orbital and the pinched waist at the middle of a p orbital. Note that at the scale of Figure 12.1 the nucleus would be an extremely small dot.

The s electron clouds are spherical: the two lowest-lying s electron clouds are shown in Figure 12.2. These s electron clouds are labeled numerically in ascending order, corresponding to increasing energy, and seem to behave vaguely like Bohr orbits by increasing their sizes as their energies increase. The analogy to planetary orbits ends when we grasp the fact that quantum mechanical electron clouds are three dimensional, whereas a Bohr orbit would be a circle lying entirely in a plane.

There is another interesting distinction between planetary orbits and quantum mechanical 1s electron clouds: s-type electron clouds have a nonzero probability at the nucleus itself! This can be seen through the phenomenon of ***electron capture***, in which a radionuclide, unstable because of an excess of nuclear protons (compared to neutrons), captures a 1s electron and thereby reduces its nuclear charge by one, gaining stability. Only an electron extremely close to the nucleus could be captured this way; 2s and 2p electrons seldom are captured, which emphasizes a point made in item 4 in the applications section at the end of Chapter 11, namely, that probability distribution ***within*** a probability cloud is not uniform, but depends on the mathematical nature of the cloud itself. After reading about the generation of X-rays in Chapter 10, you might guess correctly that electron capture, by vacating a 1s electron cloud, leads to X-ray generation in polyelectronic atoms when higher-lying electrons fall into the vacated 1s electron cloud.

The p electron clouds do not resemble planetary orbits in any way. Rather, they look a bit like a pair of eggs placed end to end, with the nucleus at the point where they touch. Note the p electron cloud carefully in Figure 12.1 — you are looking at ***one*** p electron cloud, not two. The electron has a 50% chance of being on either side of the plane (in either "egg"). This does not mean that 50% of the electron is on each side of the electron cloud — an electron is not divisible. Rather, it means that if you made repeated observations on the position of an electron in p electron cloud, half

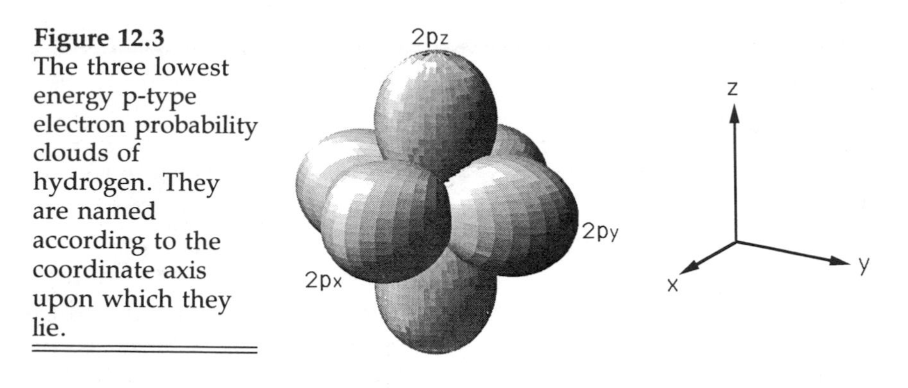

Figure 12.3
The three lowest energy p-type electron probability clouds of hydrogen. They are named according to the coordinate axis upon which they lie.

Figure 12.4
A general energy level diagram for s and p electron probability clouds. In the special case of the hydrogen atom, the 2s and 2p levels have the same energy; in all other cases, the 2p levels are higher.

2px, 2py, 2pz
2s
1s

of the time it would be on the one side of the electron cloud and half of the time it would be on the other side.

The label on p electron clouds begins with 2, not 1, and there are three 2p electron clouds at right angles to each other, as shown in Figure 12.3. Depending in the axis on which they fall, they are called $2p_x$, $2p_y$, and $2p_z$.

Using quantum mechanics it is possible to calculate the energies of electrons in the various probability clouds. A very general energy level diagram for s and p electron clouds appears in Figure 12.4. Only in the case of the hydrogen atom do the 2s and 2p electron clouds have the same energy; otherwise the 2s electron cloud has the lower energy of the two. The small circle shows that the only electron of the hydrogen atom would normally be found in the 1s electron cloud. Not all transitions between these levels are permitted and quantum mechanics correctly predicts those that are. For instance, 1s-to-2p is allowed, 1s-to-2s is forbidden, and of course no transition is allowed for energies not corresponding to actual energy level differences. Quantum mechanics also predicts the ionization energy of the hydrogen atom.

Applications, Further Discussion, and Additional Reading

1. You will find more information on quantum mechanics, presented at a moderate level of complexity, in *Fundamentals of Physics*, 2nd ed. extended, by Halliday, D. and Resnick, R., John Wiley & Sons, New York, 1981, pp. 798–813.
2. There is a macroscopic analog to the division of the probability of an electron between two lobes of a p electron cloud. A piece of paper is randomly put into one of two boxes labeled A and B. You open box A, say, and the paper is there. Upon repeating the process many times you will find that boxes A and B each will have contained the paper equally frequently. This means that there is a 50% chance of the ***entire*** piece of paper being in a given box, not that there is half a piece of paper in each box. It is the probability that is halved, not the object. The same reasoning holds for the location of an electron in the lobes of a p electron cloud.

Chapter 13
Polyelectronic Atoms

Electrons Interact with Each Other in Polyelectronic Atoms

Quantum mechanics correctly predicts the properties of the hydrogen atom, a one-electron system. We first consider helium, a two-electron atom in which there is a new consideration, namely, that the two electrons will interact with each other, as mentioned earlier in Chapter 10. Despite formidable mathematical difficulties, quantum mechanics gives a good accounting of itself in describing helium's electronic properties. We next examine some of the implications of electron-electron interactions.

Electrons can interact with each other in two ways. ***First,*** the two electrons repel one another electrostatically, which seems like a simple problem to deal with until we remember that we can no longer regard electrons in atoms as localized points. We represent their motion as probability clouds, which are mathematically complicated. Then, as the two clouds approach one another in the helium atom they distort one another electrostatically, which greatly compounds the mathematical problem of representing them. This, however, is a problem that can be addressed with some effectiveness by the use of high-speed computers.

In any case, there is very powerful evidence for the importance of electron-electron repulsion. The energy required to ionize the He^+ ion, where there can be no electron-electron repulsion, is much greater than the energy required to ionize neutral helium, where electron-electron repulsion should exist. Evidently in the case of the neutral helium, the electrons are already strongly pushing on one another, making it easier to remove one of them.

Second, electrons have an intrinsic spin, one consequence of which is that they generate magnetism, as was mentioned in Chapter 10. Thus, two electrons can attract or repel each other in accordance with the relative orientations of their north and south magnetic poles; this interaction of two magnets turns out to be of little interest to us. On the other hand, an important principle in physics forces spins to have opposite directions if they are in the same electron cloud, and ***that*** happens to be of great interest to us. We next consider these spin relationships.

The so-called ***spin direction*** is determined by a right-hand rule: point the fingers of your right hand in the direction of spin turn and your right thumb will point in the spin direction. Try it and notice how every observer derives the same spin direction no matter where she/he is standing. This is demonstrated in Figure 13.1, which you should compare with Figure 10.10.

Whenever an electrical current passes through a coil, a magnetic field is generated, as shown in Figure 13.2. Thus, a spinning electron should also act like a magnet — and it does.

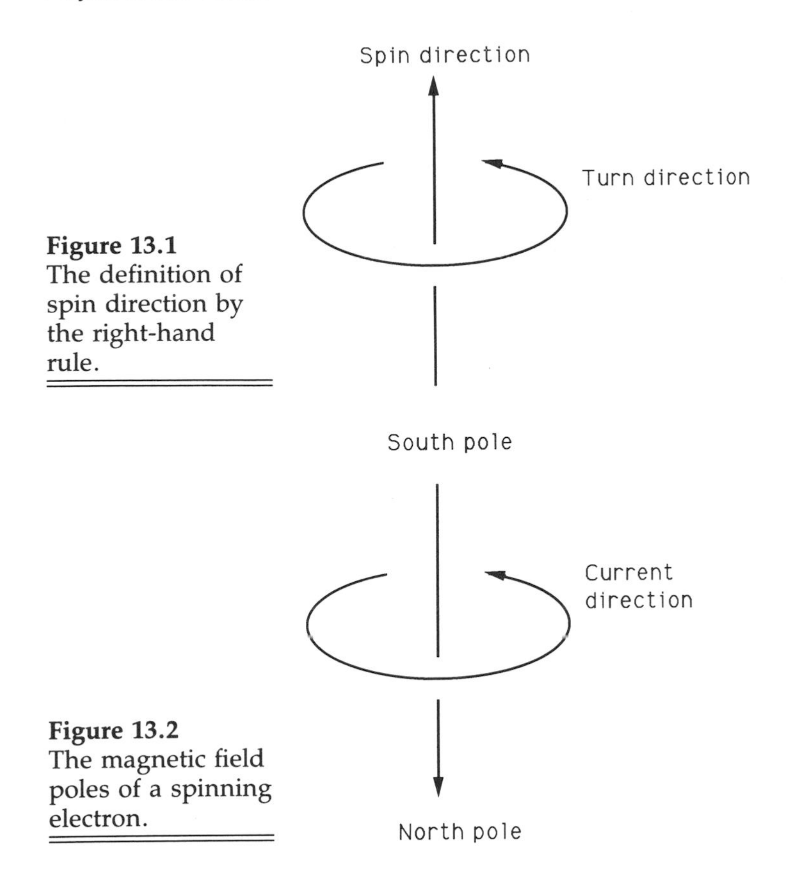

Figure 13.1
The definition of spin direction by the right-hand rule.

Figure 13.2
The magnetic field poles of a spinning electron.

A peculiar thing about electron spin is that quantum mechanics permits only two spin directions which we will call "up" and "down". This brings up a new question: what do "up" and "down" mean to an electron? Gravity barely affects lightweight objects like electrons, so that won't do for a defining parameter. Instead, we take a large bar magnet and place the spinning electron between the bars. The small electron magnet will then align itself spin up or spin down, thus defining the notions of "up" and "down", in terms of the bar magnet (see Figure 13.3). You should note the important role played by the macroscopic bar magnet in the spinning-electron model.

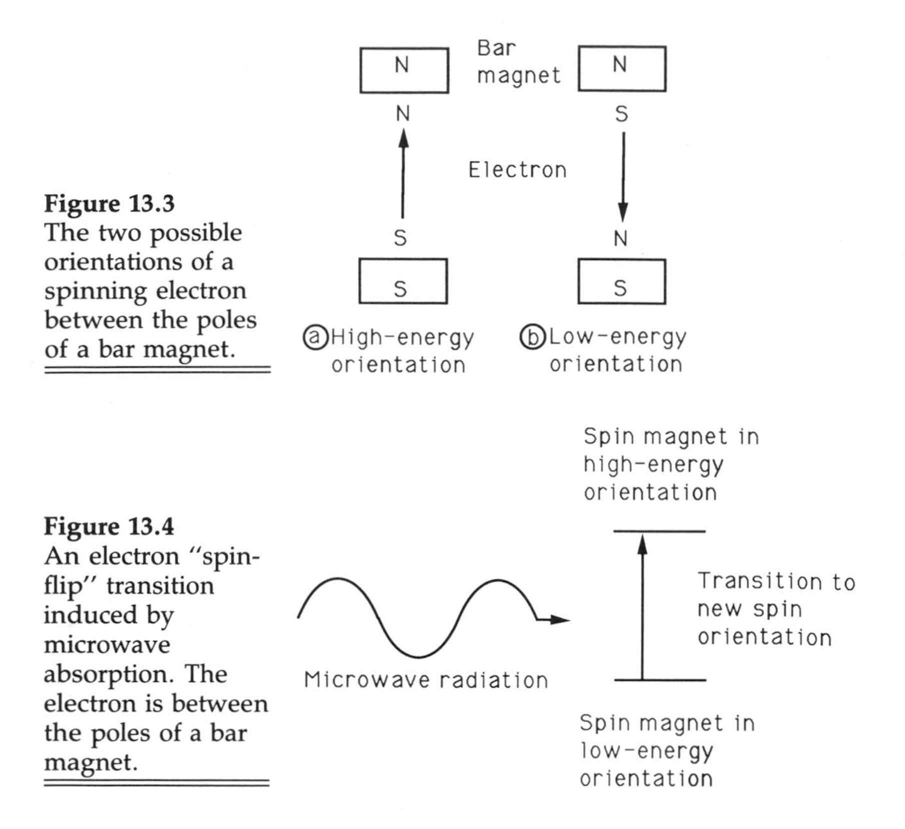

Figure 13.3 The two possible orientations of a spinning electron between the poles of a bar magnet.

Figure 13.4 An electron "spin-flip" transition induced by microwave absorption. The electron is between the poles of a bar magnet.

The orientation of Figure 13.3a has a higher energy than that of 13.3b. We expect this because of the north-north and south-south conjunctions of the poles of the electron and the bar magnet, which should be unstable. Absorption of radiation with energy of about 10^{-5} eV (called "microwaves") can cause an electron to change its spin from the low energy orientation to that of the high energy orientation, as shown in Figure 13.4. The excited spin state will later fall back to the ground spin state and emit the excess energy.

The exact energy difference between the two spin states depends on the structure of the molecule in the region of the unpaired spin. The method of ***electron spin resonance*** capitalizes on this property by correlating structural properties with the exact microwave energy absorbed by the particle with the unpaired spin, such a particle being called a ***radical***.

Getting back to helium, a very deep principle in physics dictates that two electrons in the same electron cloud must have opposite spins. In other words, their overall physical states must differ: if their electron cloud shapes (space states) are the same, then their spin states must be different. This condition can be met for ground-state helium by putting both electrons into the 1s electron cloud, but giving them ***opposite spins.*** Thus, the electrons will have the same configuration in space, but different spin configurations, yielding different overall physical states.

Using this general notion, called the Exclusion Principle, we can build other atoms. The simplest way to do it is to use the hydrogen energy-level diagram, arranging s and p levels in order of increasing energy, and then filling the levels upward with electrons, making sure that no space electron cloud has more than two electrons. If there are two electrons in one electron cloud then they must have opposite (antiparallel) spins. Figure 13.5 shows the various energy levels of a neon atom constructed in this way.

Figure 13.6 shows an electron cloud picture corresponding to the energy-level picture of Figure 13.5. Each of the various orbitals of neon — 1s, 2s, $2p_x$, $2p_y$, and $2p_z$ — would contain two electrons. We will henceforth use hydrogen probability clouds in this way for atoms other than hydrogen, but we must be careful. First, as was discussed in conjunction with Figures 10.8 and 10.9, electrons near heavier nuclei are held more tightly than

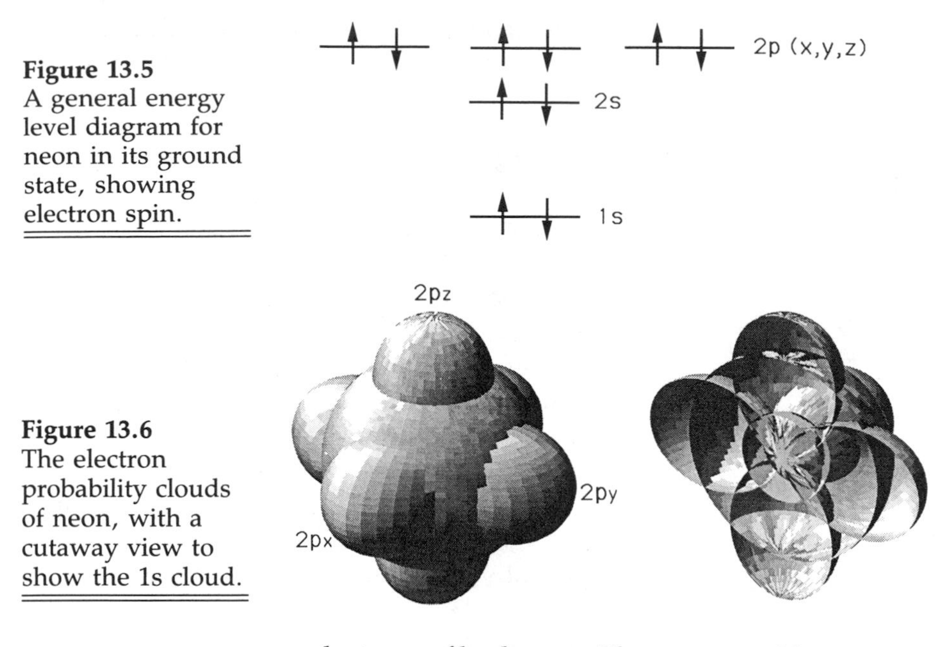

Figure 13.5
A general energy level diagram for neon in its ground state, showing electron spin.

Figure 13.6
The electron probability clouds of neon, with a cutaway view to show the 1s cloud.

electrons of hydrogen. Thus, we would expect the various charge clouds of heavier atoms to be smaller than their counterparts in hydrogen. Second, in filling the 2p clouds we first put one electron into ***each*** of the 2p clouds and then go back and put in a second one, if needed. Thus, we "construct" the electronic configurations of carbon, nitrogen, and oxygen atoms in the following way

	Number of electrons				
	1s	**2s**	**$2p_x$**	**$2p_y$**	**$2p_z$**
Carbon	2	2	1	1	0
Nitrogen	2	2	1	1	1
Oxygen	2	2	2	1	1

The reason, for example, for not immediately putting two electrons into a $2p_x$ cloud of carbon is that such an arrangement would increase repulsion between those two electrons. Much less repulsion results from putting one each into the $2p_x$ and $2p_y$ clouds. Of course, we have no choice in the case of oxygen, we have run out of p clouds into which to partition the electrons and are forced to put two into the $2p_x$ cloud. (Remember that if a cloud contains two electrons they must have opposite spins.)

In concluding this chapter we should note several things. ***First,*** a sophisticated treatment of polyelectronic atoms includes explicit consideration of the electrostatic repulsion between electrons. ***Second,*** the antiparallel spin requirement for two electrons in a single space electron cloud is ***not*** the result of the north and south poles of the two spin magnets attracting each other — that is a relatively unimportant interaction. Rather, the antiparallel spin requirement is the result of the electrons taking the lowest available energy level and then the Exclusion Principle requiring that electrons have different overall states. This latter condition dictates that, if two electrons have the same space state, then they must have different spin states, making their respective overall states different. Clearly, two 1s helium electrons, with opposite spins, fit this description. ***Third,*** as mentioned earlier, there are electron clouds other than just the s and p types to which we are restricting ourselves. No considerations important to us are lost by omitting the others.

Applications, Further Discussion, and Additional Reading

1. Let us consider the matter of determining the spin direction of an electron in the absence of the external (bar magnet) magnetic field. After all, if it is the external field that determines up and down, what might we say about the spin direction when the external magnetic field is

turned off? Actually, we can say ***nothing.*** The external manifestation of electron spin is magnetism and we must communicate with it on that basis — using another, external magnet. However, it is the external magnet that forces the electron into one of the two orientations in Figure 13.3. We might imagine that there is some notion of spin orientation in the absence of the external magnet, but we cannot know it because we ***must*** activate the external magnet to interact with electron spin and that forces the electron into one of the allowed spin states, assuming it is not already in one. Our act of observation creates the condition!

Chapter 14
The Covalent Bond

Electrons Form Probability Clouds Around Molecules

The elements that are most common in living systems are also common in nonliving systems; these elements are hydrogen, carbon, nitrogen, and oxygen. Thus, the essential properties of living systems are not to be found at the atomic level. Rather, those properties are to be understood in terms of ***organization*** at many levels — molecular, supramolecular, organelle, and cellular. This chapter is an introduction to "organization" at the molecular level.

There are models for molecular structure, just as there are for atoms, and these models can be as complicated as we care to make them. If we were interested in accurate predictions of numerical data, complex models would be required. Our interests, however, are in qualitative behavior and we are free to omit many refinements that a student in an advanced course might need. Nevertheless, our simple models will predict most of the important structural features of molecules found in living systems.

Consider two hydrogen atoms, as shown in Figure 14.1a. The atoms are very far apart and do not interact electrostatically, their

(b)

Superposition; intense repulsion

Figure 14.1
Two hydrogen atoms at varying separatory distances.

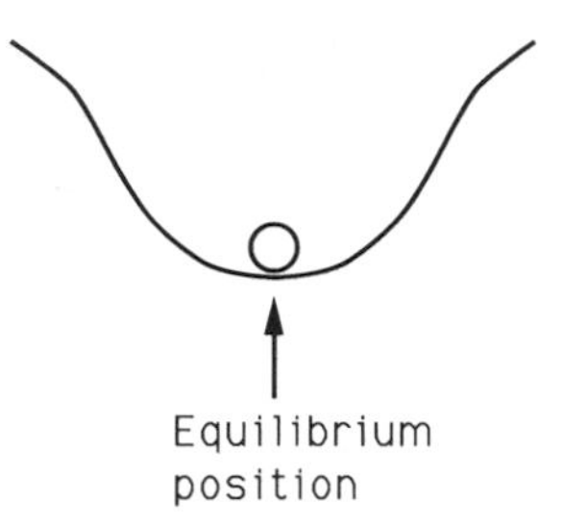

Figure 14.2
Mechanical equilibrium. Movement of the ball left or right goes uphill, therefore requires energy.

interaction is zero. In Figure 14.1b the two atoms are virtually superimposed and repel one another strongly. There is an intermediate arrangement, with the nuclei about 0.1 nm apart, in which their orientation is stable, meaning that energy would be required to move the nuclei ***either*** further apart ***or*** closer together (see Figure 14.1c). We will say that a bond has formed between the two atoms, creating the diatomic molecule H_2.

There is a simple macroscopic analog which demonstrates the concept of equilibrium distance, shown in Figure 14.2. A marble comes to rest at the bottom of a two-dimensional cup because to move in either direction, left or right, would require the input of energy (to go uphill). It is similar with the arrangement

Figure 14.3
A covalent bond formed by the overlap of the 1s probability clouds of two hydrogen atoms. The electron sharing occurs in the overlap region shown in the cutaway view.

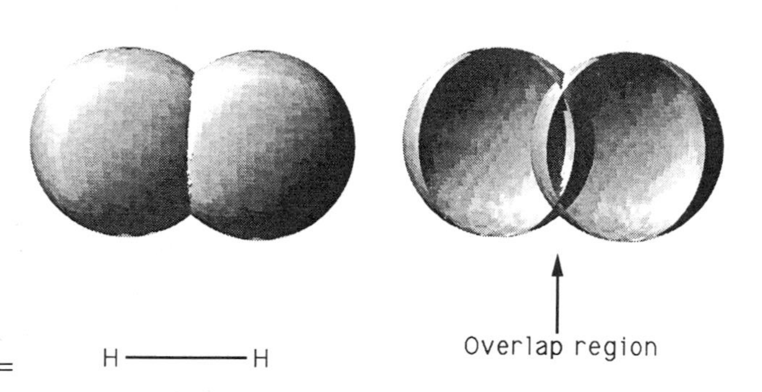

of hydrogen atoms in Figure 14.1c; moving the atoms either closer together or further apart would require energy. Thus, there is a stable separation distance.

The nature of the attraction between the two atoms in Figure 14.1c is just of the ordinary electrostatic sort, but having said that, we feel that the probability clouds involved must be different from the atomic case. Otherwise, what would distinguish a diatomic molecule from two ordinary atoms, sitting side by side?

To answer this question let us again imagine two hydrogen atoms at a large distance apart and then move them toward each other. As they get close we find that their atomic electron clouds combine to form ***molecular electron probability clouds.*** Figure 14.3 shows a molecular electron probability cloud formed by the two hydrogen 1s atomic electron clouds.

There are several important points to be made about this molecular electron cloud:

1. The molecular electron cloud has a circular cross section along the line connecting the nuclei. It is said to have ***axial symmetry*** and is called a sigma (σ) molecular electron cloud.

2. This molecular electron cloud contains the two original atomic electrons, with antiparallel spins. Occupancy for molecular electron clouds is dictated by the Exclusion Principle, meaning that there may be zero, one, or two electrons in such a cloud, but they must have opposite spins if there are two. The energetically most stable arrangement of electrons constitutes the ***molecular ground state.***
3. Potentially, there are other molecular electron clouds at energies above the one in Figure 14.3 (but they are not shown). A ground state electron may be excited to an excited state if a quantum of the correct energy is absorbed (just like in the case of atomic electrons).
4. By absorbing the correct energy, this molecule can be ionized, just like in the atomic case.
5. The two electrons spend most of their time in the region ***between*** the two nuclei. In other words, that is the molecular electron cloud's region of highest electron probability density. This high central density is responsible for holding the molecule together because ***each*** electron is electrostatically attracted to ***both*** nuclei. The electrons are thus said to be shared, and that sharing is the "glue" holding the molecule together. In fact, this electron sharing is the essence of the covalent bond (see next paragraph).

You may be wondering about this fifth point — after all, if the two electrons spend most of their time in the same region between the nuclei, why don't they electrostatically repel one another and move to different regions, outside the nuclei, thus breaking the bond? In fact, the two electrons *do* repel one another as expected, but there is another factor to be reckoned with, one that has no macroscopic analog and which tends to keep the electrons in the same region of space in spite of electrostatic repulsion. We will call this factor the "exchange factor": electrons are indistinguishable from one another and that fact is built into their quantum mechanical description by acknowledging that the two electrons could exchange places with each other without changing the shape of the probability cloud of the molecular ground state. The two electrons must be physically close together to make this exchange of places feasible. Thus, the exchange factor causes a partial confinement of the two electrons to the internuclear region of the molecule and the Exclusion Principle causes their spins to be paired (antiparallel). This bond, described by the sharing of two electrons of opposite spin by two nuclei, is called a ***covalent bond.***

Covalent bonds are quite strong, as chemical bonds go, requiring anywhere from about 3 to 8 eV for breakage, depending on the atoms involved. We can now understand why pure hydrogen almost invariably exists in nature as the diatomic molecule H_2, not as the atom H. Atomic hydrogen is a radical — a particle having an unpaired electron spin. Any radical would quickly react with another one to form a (strong) covalent bond. Thus, radicals such as atomic hydrogen do not last very long in nature — they're too reactive. On the other hand, we also see why ordinary helium is unreactive: its electrons are already spin-paired.

Figure 14.4
A schematic picture of a covalent bond between a carbon and a hydrogen atom, showing the spin-paired electrons and omitting the probability clouds.

C ↑↓ H

Also written, C —— H

Carbon Forms Hybrid Orbitals

We have combined two hydrogen atoms into a covalently bonded H_2 molecule, in which electron spin pairing is a major consideration. Now we can use the same approach to combining a hydrogen atom and a carbon atom, as shown schematically in Figure 14.4. From this point on, all the covalent bonds to be described will involve the concept of spin pairing shown in that figure.

The two unpaired spins of atomic carbon are in two 2p electron clouds, from which we could ostensibly get spin pairing to yield the molecule CH_2. Unfortunately, the least number of hydrogen atoms to which one carbon atom stably bonds is four, not two. The compound CH_4 is methane and any description of its molecular electron clouds must explain its properties.

It is hypothesized that when carbon covalently bonds, its electron clouds change from free atomic forms to so-called ***hybrid*** forms, whose existence only makes sense in terms of molecular chemical bonds, not in terms of atoms. We can justify such an *ad hoc* assumption on two grounds. First, it leads to accurate predictions and, second, probability clouds will certainly change as they approach one another to form a bond. Cloud shapes should be strongly modified by the formation of the molecule.

Descriptions of these modified charge clouds can be mathematically horrendous, even in an

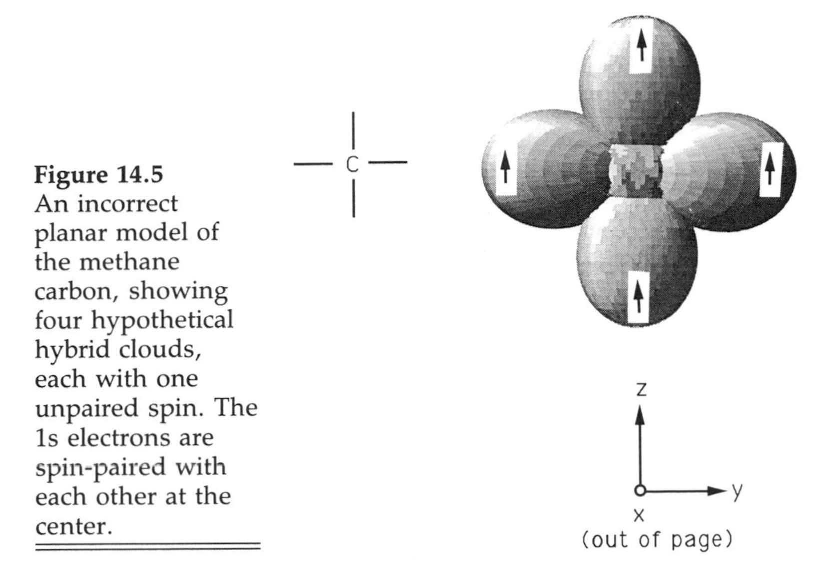

Figure 14.5
An incorrect planar model of the methane carbon, showing four hypothetical hybrid clouds, each with one unpaired spin. The 1s electrons are spin-paired with each other at the center.

era of computers. Chemists and physicists approach the problem by representing these complicated functions as combinations of known, simpler probability functions, namely, atomic probability functions. These arithmetic combinations are called ***hybrid electron clouds*** and they represent a tractable best estimate as to what shape the atom takes when bonding.

We can conclude that the four hybrid electron clouds of the methane carbon probably involve the two 2s and the two 2p electrons, because being furthest from the nucleus those are the least tightly held, and so they are the most available for bonding. (They are called ***valence electrons.***) Thus, four hybrid electron clouds are hybridized by combining the 2s and 2p wavefunctions (orbitals) and squaring the result, as described in the third section of Chapter 11. Without doing the calculations we might guess that a possible result would be like that of Figure 14.5, which accounts for all the carbon electrons. It is, however, incorrect because it is planar and methane is not.

We can understand the problem easily enough by considering Figure 14.5, which shows the (incorrect) planar model of a methane carbon. Molecules can rotate in space, in a tumbling motion. It is clear that there are two unique kinds of rotation possible for methane: around the y or the z axes, which are identical, ***and*** a different one around the x axis. It may help to picture the (incorrect) methane carbon of Figure 14.5 as fan blades, with the motor axle along the x axis and the y and z axes along the lengths of the fan blades, at right angles to one another. You would not be able to distinguish rotation about the y and z axes from one another, but rotation about the x axis is clearly different from the other two. In this model methane should have two distinguishable kinds of rotation, one twice as common as the other. In fact, however, methane experimentally shows only one pattern of rotation.

Possession of only a single kind of rotation is characteristic of a sphere because the rotation of a sphere is the same about every axis; thus, the hybridization of methane must approximate a sphere. Figure 14.6A and Figure 14.6B show the accepted hybridization scheme for a methane carbon, one which confers a spherical shape in the sense that there is no way to distinguish any one direction from any other. This conclusion specifically follows from Figure 14.6B, in which it is shown that all of the hybrid clouds are at 109° from one another and have identical appearances.

The hybrid structure described in Figure 14.6A and Figure 14.6B is said to be ***tetrahedral*** and is symbolized as sp^3 because it combines one 2s wavefunction with all three 2p wavefunctions (recall that the square of the wavefunction gives the probability cloud).

One sp^3 hybrid cloud

All four sp^3 clouds

A cutaway of the four clouds

Carbon 1s electrons are omitted

Figure 14.6A
A hybridization model (sp^3) of methane which is consistent with experiments.

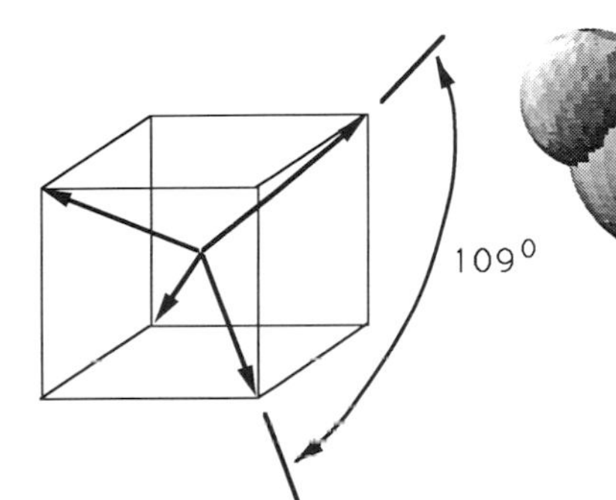

Simplified view of the sp^3 clouds, showing bond angles, all of which are 109°

The complete methane molecule

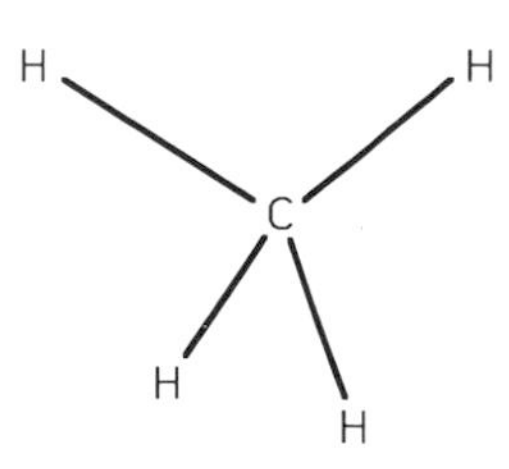

Figure 14.6B
A hybridization model (sp^3) of methane which is consistent with experiments, continued.

There are two important ancillary points about Figure 14.6A and Figure 14.6B. First, the 1s electrons will usually be omitted from now on

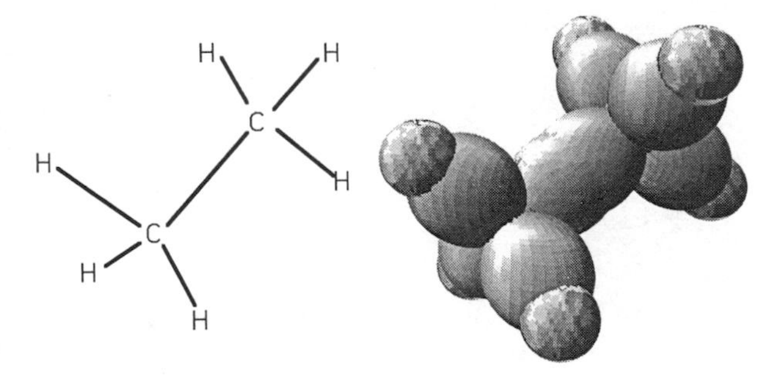

Figure 14.7
The valence electron clouds of ethane. The hybrid clouds of the carbons are sp^3.

because they are not valence electrons and do not participate in hybridization; they therefore tend to clutter up our diagrams. Second, the resemblance between these hybrid electron clouds and 2p electron clouds is only superficial; they are quite different. There are three 2p electron clouds, each with two equal lobes, one on each side of the nucleus, as shown in Figure 12.3. Each of the four hybrid electron clouds has virtually all of its probability on only one side of the nucleus, as seen in Figure 14.6A. The nucleus is at the pinched waist.

The C–H covalent bonds shown in Figure 14.6B have circular cross sections perpendicular to the internuclear axis; they are σ bonds because they have axial symmetry.

Figure 14.7 shows the electron clouds of the valence electrons of the carbon atoms of ethane ($H_3C–CH_3$); the C–H bonds are identical to those of methane. The C–C bond involves two carbon sp^3 hybrid orbitals in a σ-type bond. It should be clear that carbon atoms could be strung together this way into even more complex structures. The capacity of carbon to be bonded into an unlimited range of molecular arrangements makes it especially suitable for constructing living systems. For example, the carbon-to-carbon σ bonds of Figure 14.7 are found in hydrocarbons, amino acids, lipids,

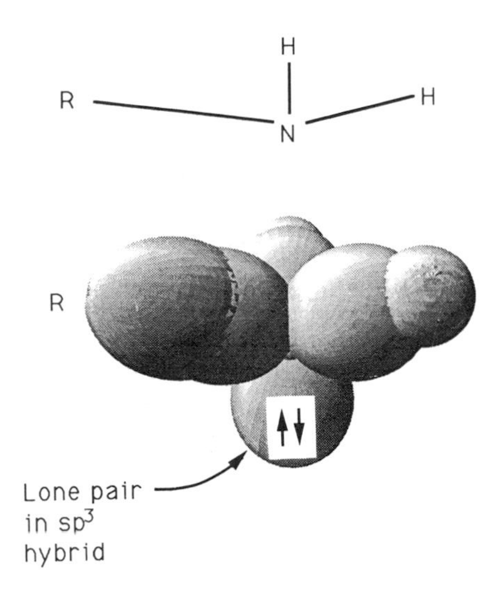

Figure 14.8
The valence electron clouds of the nitrogen amino group. The hybrid clouds of the nitrogen are sp^3 and there is a lone pair.

polysaccharides, and nucleotides, to name a few.

Before moving on to hybrids of other atoms it is important to reemphasize that there is nothing magical about the notion of orbital hybridization. It is merely a way of using known, mathematically tractable atomic wavefunctions to describe the properties of an atom as it enters into molecular bonds, this being a physical process which is certain to change the original electron clouds. In addition to its computational convenience, hybridization provides accurate descriptions of molecular structure and behavior. In earlier chapters we saw that even if there were some hidden "underlying reality" at the submicroscopic level we would still be forced to use a macroscopic world model to describe it, and that's exactly what the hybrid electron cloud model does.

Nitrogen and Oxygen Can Form Hybrids

Figure 14.8 shows the nitrogen amino group. The symbol "R" (for ***radical***) represents any group capable of entering into a covalent bond, e.g., hydrogen, carbon, an ethyl group,

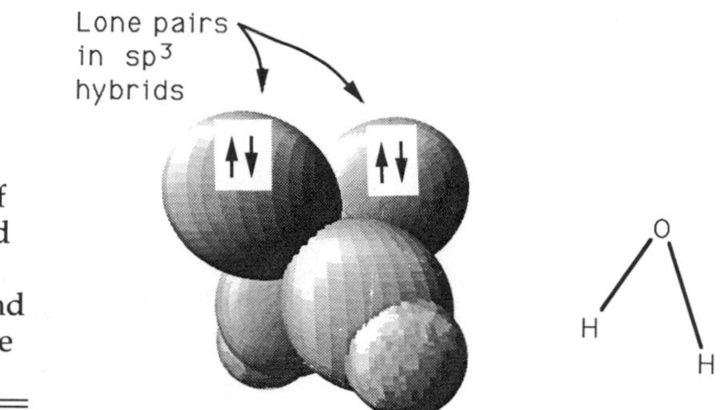

Figure 14.9
The valence electron clouds of water. The hybrid clouds of the oxygen are sp^3 and there are two lone pairs.

whatever. The hybridization of the amino nitrogen is sp^3 (tetragonal) and we see that two of the nitrogen's seven electrons spin-pair among themselves, forming what is called a "lone pair". Thus, the amino nitrogen forms three covalent bonds and a lone pair, the remaining two electrons of the seven being in the 1s electron cloud. The covalent bonds from the amino nitrogen are all of the axial σ type. The amino group of nitrogen is found in amino acids, ammonia, and nucleotides.

Figure 14.9 shows the hybridization of the hydroxyl oxygen, typified by water, to be sp^3. There are two electrons in the 1s electron cloud, two electrons in the hybrid electron clouds participate in covalent bonds, and there are ***two*** lone pairs. Note the similarity of the oxygen hybrids to those of carbon and nitrogen in Figures 14.7 and 14.8. As before, the sp^3 electron clouds are ***approximately*** equivalent and tetragonal, although we might guess (correctly) that the two lone pairs push at each other very hard, somewhat increasing the angle of separation. We can see that the water molecule is not linear, but is bent at the oxygen into a nominally 109° angle. The O–H bonds are of the σ type; this group is found in alcohols and organic acids, for example. A similar "–O–" structure is found in esters.

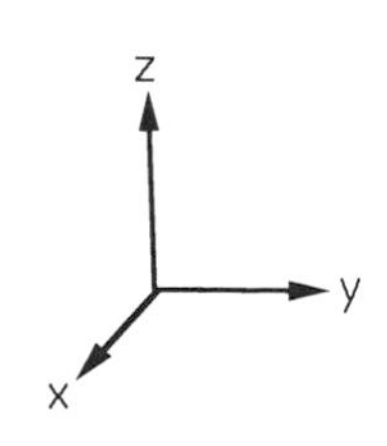

View from just above the xy-plane

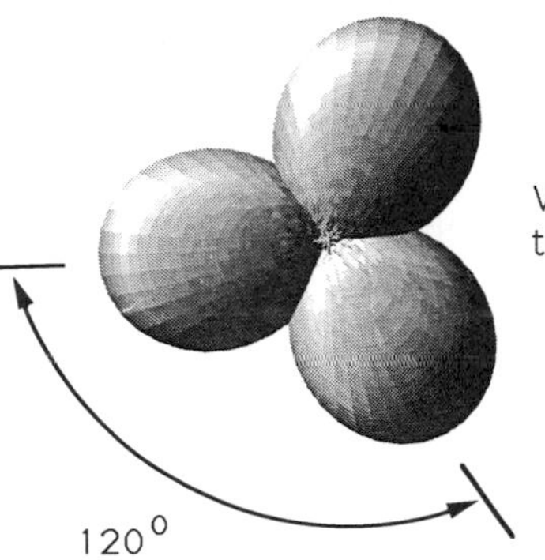

View from along the z-axis

Figure 14.10 Two views of the sp^2 hybrid clouds of carbon. All three clouds lie in a plane.

Carbon, Nitrogen, and Oxygen Can Form Double Bonds

Carbon can hybridize in another way than sp^3. The 2s electron cloud and two 2p electron clouds can form an sp^2 hybrid, in which there are three hybrids at 120° from each other and lying in a plane (see Figure 14.10).

In fact, Figure 14.10 is incomplete because it only accounts for five electrons: two in the 1s cloud and one in each of the three hybrid clouds. Figure 14.11 includes the remaining electron: there are two 1s electrons, three electrons in the carbon sp^2 hybrid electron clouds, and one $2p_z$ electron left over. Thus, the valence electrons of this carbon atom are a combination of atomic $2p_z$ and hybrid sp^2 electron clouds.

We now construct ethene ($H_2C{-}CH_2$), as shown in Figure 14.12. Compare it with ethane

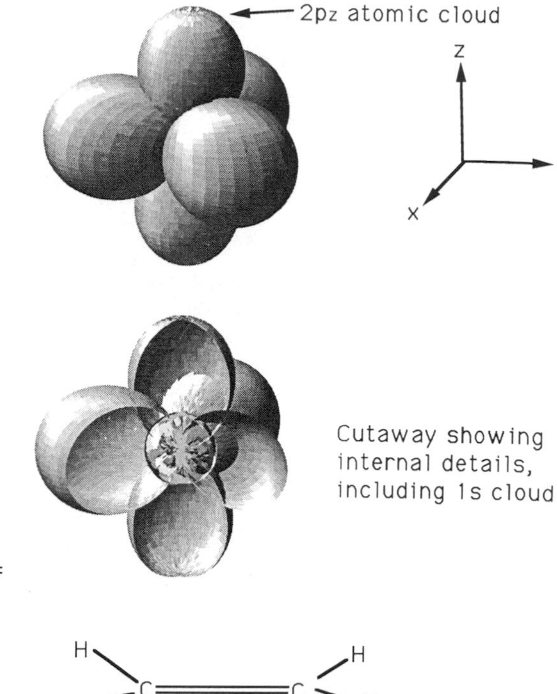

Figure 14.11 The sp^2 hybrid clouds of carbon, with the $2p_z$ atomic cloud shown.

Figure 14.12 The valence electron clouds of double-bonded carbon in ethene. The $2p_z$ atomic clouds form the second bond and the hybrid clouds are sp^2.

in Figure 14.7. Each carbon of ethene has two σ-type covalent bonds to hydrogen and another to the opposite carbon. The new wrinkle here is the bond formed by spin pairing between the electrons of the atomic $2p_z$ clouds, whose orientation is perpendicular to the plane of the other electron clouds. At the scale of the figure, the overlap of the two $2p_z$ clouds is not obvious, but it is there nevertheless. Be very careful about interpreting this bond: there

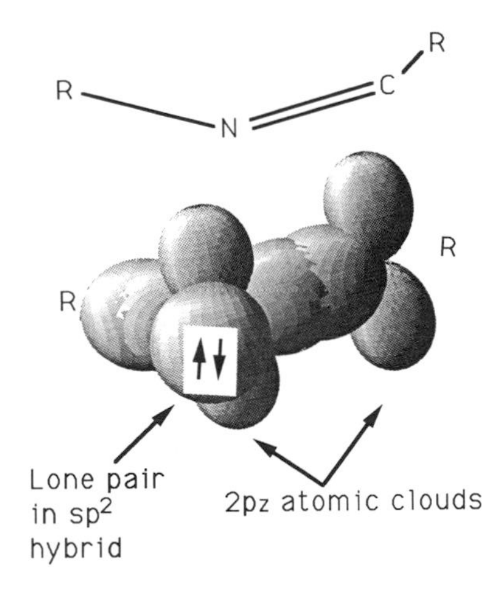

Figure 14.13 The valence electron clouds of double bonded nitrogen. The $2p_z$ atomic clouds form the second bond, the hybrid clouds are sp^2 and there is a lone pair.

is only ***one*** bond formed by the two $2p_z$ electron clouds because they each contain only one electron, as explained in Chapter 13. Half the probability of the p electrons is above the molecular plane and half is below. There is, however, a ***total*** of two bonds linking the carbons — one through the sp^2 hybrids and one from the p_z electrons. The molecular electron cloud formed by the two p_z atomic electron clouds above and below the plane of the molecule has mirror symmetry (reflection in the plane) and is called a π (pi) molecular electron cloud. We say that the carbons are connected by a ***double bond***, one of which is a σ type and the other is a π type. Double-bonded carbon atoms are found in fatty acids of cell membranes, for example.

Nitrogen can form sp^2 hybrids and double bonds, as shown in Figure 14.13. Note the lone pair. Double-bonded nitrogen atoms are found in nucleotides of DNA and RNA, for example.

Oxygen can form sp^2 hybrids and double bonds as shown in Figure 14.14. Note the lone pairs. Double-bonded oxygen atoms are found in ketones, aldehydes, and organic acids.

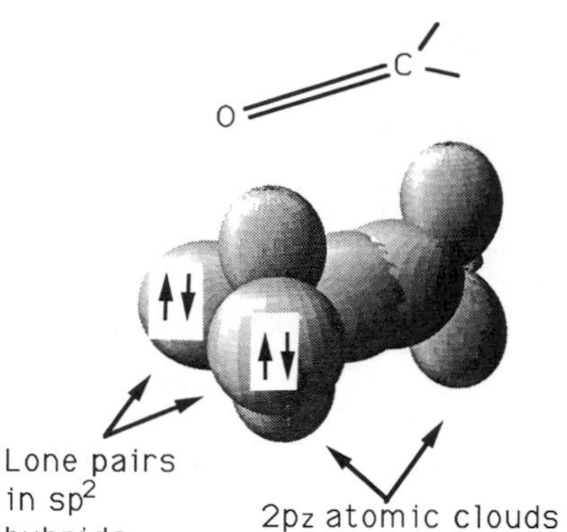

Figure 14.14
The valence electron clouds of double-bonded oxygen. The $2p_z$ atomic clouds form the second bond, the hybrids are sp^2 and there are two lone pairs.

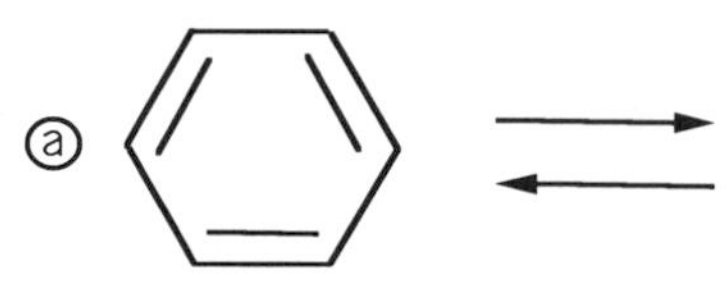

Figure 14.15
Benzene, showing equivalent structures which are the origin of partial double bonds.

The structures of most of the molecules of interest in biology can be assembled from the atoms and hybridization schemes presented in this chapter. There will of course be special situations, one of which is the case of benzene, where the bonds seem to be part single and part double. Figure 14.15a shows two equivalent pictures of benzene, "equivalent" because they are both correct at the same time. In either case there are 9 (1 + 2 + 1 + 2 + 1 + 2) carbon-carbon bonds, or an average of 1.5 bonds between each pair of carbon atoms. We can therefore represent benzene as a kind of intermediate between the two pictures, represented in Figure 14.15b, where ***each of the six*** of the carbon-carbon linkages seems to contain 1.5 bonds, meaning that they are

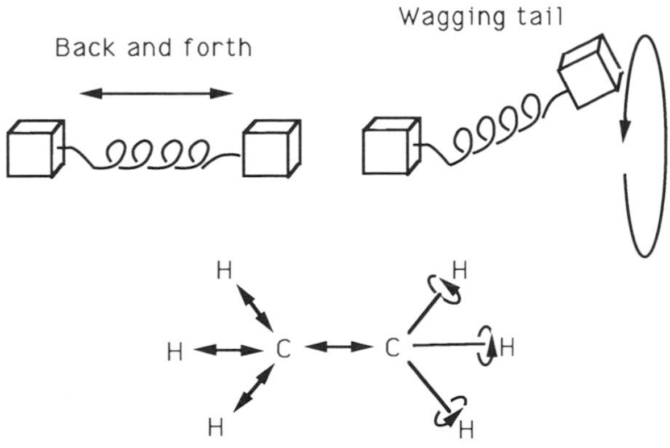

Figure 14.16 Example of vibrations associated with a chemical bond, showing two mechanical analogs and ethane. The IR energies absorbed by a bond are unique for each atomic pair forming the bond.

halfway between a single and a double bond, or have 50% double-bonded character. The macroscopic world approximation is very obvious in the notion of "1.5 bonds". (The actual number calculated by theoretical chemists is 1.67 bonds, but the basic idea is the same.)

Nuclei Vibrate Across Bonds

This chapter presents us with the opportunity to study the absorption and emission of another kind of radiation: infrared (IR). Two nuclei at the ends of a bond are not stationary; rather, they seem to vibrate back and forth about some average position between them, like weights on the ends of a spring, as shown in Figure 14.16.

Only ***specific*** changes in the energy of molecular vibrations are permitted, because that energy is quantized; in fact, bond vibrational energies can be represented by an energy level diagram similar to those associated with electrons. If light of the correct IR energy (a few tenths of an electron volt) is absorbed, the two nuclei will adopt a higher vibrational energy; on the other hand, they may also emit an IR quantum and then adopt a lower vibrational energy. The actual permitted energies are determined by the natures of the nuclei at the ends of the bond and by the nature of the bond itself, i.e., covalent or H bond. Being so specific, the IR frequencies absorbed and emitted by an unknown compound are very useful in identifying its component parts.

As mentioned in Chapter 8, nuclear isotopic substitution at one or both ends of a covalent bond should affect the spacing between vibrational energy levels of the bond. Thus, the IR absorption and emission spectra of a C–H bond should be different from that of a C–D bond because the deuteron is more massive than the proton.

APPLICATIONS, FURTHER DISCUSSION, AND ADDITIONAL READING

1. A readable discussion of molecular shapes can be found in *Geometry of Molecules*, by Price, C. C., McGraw-Hill, New York, 1971.
2. The symbols σ and π have been used in the text above to describe the shapes of both covalent bonds and the molecular electron clouds that constitute those bonds. This should introduce no ambiguity as long as there is an actual bond formed by the molecular electron clouds. There ***are***, however, excited state molecular electron clouds which are ***antibonding***, i.e., they lead to dissociation of the molecule. The shapes of these antibonding molecular electron clouds can still be described as σ or π, but, being dissociative, the idea of a corresponding covalent bond is lacking.
3. An extensive, but nonmathematical, treatment of covalent bonding is contained in *Chemical Bonding Clarified Through Quantum Mechanics*, by Pimental, G. C. and Spratley, R. D., Holden-Day, San Francisco, 1969.

4. The nuclei at the ends of a bond seem to vibrate about an average position. To see this, refer back to Figure 14.16, which is a macroscopic model of a covalent bond. If the spring were at rest, it would adopt some stable equilibrium separation of the two weights; when energized, the spring vibrates back and forth about this position. Covalent (and all other) bonds are represented in the same way. Thus, when a bond length is quoted in the scientific literature it is this average distance that is intended.

5. The process of replacing a complicated probability function with an arithmetic combination of simpler functions has a macroscopic analog. Suppose you build an engine that runs best on a pure fuel which, unfortunately, is unobtainable. Instead, you use a substitute fuel that is a blend of easily obtained compounds on which the engine runs well. That is the thinking behind the construction of hybrid electron clouds: substitute a blend of (less-complicated) atomic electron clouds for a (complicated) molecular electron cloud.

6. In Chapter 13 you were introduced to the notion of a radical, a particle with an unpaired electron spin. Now, after reading Chapter 14 you should be able to understand why stable radicals are rare in nature: two radicals will quickly form a covalent bond, eliminating both unpaired electron spins by pairing them. Suppose there were two separate functional biological molecules and X-rays ionized each, leaving each as a radical. A covalent bond could then bridge the two molecules,

forming a cross-link, and thereby alter the function of ***both*** molecules. This destructive covalent cross-linking is a commonly observed effect of ionizing radiation.

7. As an exercise, determine the hybridization of the atoms in the following compounds: $CH_3{-}CH_2{-}CH_2{-}CH_3$ (butane), $CH_3{-}CH_2OH$ (ethyl alcohol, where the $-OH$ group looks like that of water), $CH_2{=}CH{-}CH{=}CH_2$ (butadiene),

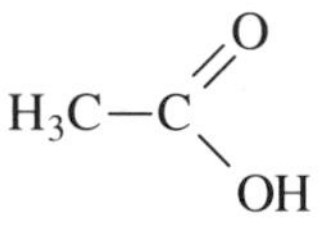

   O
   H_3C-C
   OH
   (acetic acid)

   and

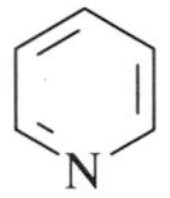

   N
   (pyridine)

8. Perhaps the notion that a charge cloud can "contain" zero electrons strikes you as peculiar; in fact, it is a common concept in physics. For example, you might think of a circular orbit at a distance of 100 miles from the earth. The abstract idea of the orbit exists even if there is no satellite present. Thus, we can imagine the shape of a charge cloud even if there is no electron in it.

9. The methane molecule is not a perfect sphere, in the sense of a globe. Rather, as pointed out above, the lobes of the hybrids are ***all*** at 109° from one another and they look alike. Thus, there is no way to distinguish any one direction from any other one. It is in this sense that methane has spherical symmetry.

10. One effect of a double bond in, say, ethene, is to prevent the two carbons from rotating with respect to each other. The reason is that the π electron clouds define a unique plane through the double bond. The effect of this is to cause all four carbon-to-hydrogen bonds to lie in a single plane. Note that this restriction of rotation does not apply to ethane because a σ bond does not define a plane (it is axial and looks the same no matter how the two carbons are rotated with respect to each other).
11. Examples were given above of organic molecules possessing the various atoms and their hybrid electron clouds. You can find the structural formulas for these molecules in any biochemistry book. For example, a good reference is *Biochemistry,* 2nd ed., by Stryer, L., W. H. Freeman, San Francisco, 1981.

Chapter 15
The Ionic Bond

Some Bonds Are Highly Asymmetrical

Properly said, a ***pure*** covalent bond can exist only between identical nuclei, e.g., two carbon nuclei bonded together. If the nuclei are different, the center of charge in the bond always moves closer to one of the two nuclei, conferring electrical polarity on the bond; this polarity gives the bond extra stability. We say that the bond has an ***ionic*** character in addition to its covalent character.

We should expect charge asymmetry in the heteronuclear case because the different atoms have completely different electronic structures and nuclear compositions and should therefore exert completely different forces on the shared electrons of a bond. For example, higher atomic number nuclei, with a greater positive charge, should exert a greater attraction on electrons in bonds; on the other hand, higher atomic number nuclei also have more inner electrons, which will shield electrons in bonds from the nucleus. Thus, bond asymmetry will be a compromise between the two effects.

Chemists have coined the word ''electronegativity'' to describe the degree to which a given atom seems to attract the valence

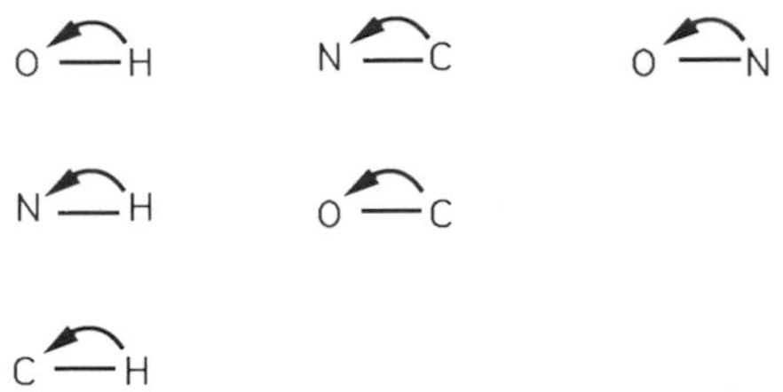

Figure 15.1
The shift of electrons across bonds toward atoms of greater electronegativity.

electrons toward itself. They have shown that the increasing order of electronegativity is H, C, N, O (see Figure 15.1).

In the figure the arrows point to the atom that holds the greater share of the charge of the bond. For example, in the case of H–C shown in Figure 15.1, the carbon attracts the bond electrons more than does the hydrogen, causing a slight negative charge on the C and a slight positive charge on the H.

In the most extreme case the electron from one nucleus becomes completely "captured" by the other nucleus; when this happens the "sharing of an electron pair" is no longer a reasonable explanation for the bond — rather, the bonding is better described as the attraction between positive and negative ions. The strengths of these ***ionic*** bonds are about the same as those of pure covalent bonds.

Distinctly ionic bonds, with little covalent character, are much more common in inorganic compounds (e.g., sodium chloride — table salt) than in biochemicals and we will not need to be too concerned about them. We acknowledge that there is some ionic ***contribution*** in all heteronuclear covalent bonds and if that contribution should become an important consideration we will defer to it. In fact, we will do that in the next chapter.

APPLICATIONS, FURTHER DISCUSSION, AND ADDITIONAL READING

1. An extensive, but nonmathematical, treatment of ionic bonding is contained in *Chemical Bonding Clarified Through Quantum Mechanics,* by Pimentel, G. C. and Spratley, R. D., Holden-Day, San Francisco, 1969.

Chapter 16
The Hydrogen Bond

Lone Pairs and Hydrogen Protons Attract One Another

Consider the valence electron distribution of water in Figure 16.1. Water is dipolar, the negative end being at the lone pairs. On the basis of the discussion of bonds in the previous chapters, we can now see the source of the positive end. The electrons in the covalent bonds between oxygen and hydrogen are skewed toward the more electronegative oxygen, leaving the hydrogen proton somewhat exposed (see Figure 16.2).

There is an attraction between a lone pair of one water molecule and a hydrogen proton of another water molecule and that interaction is called a ***hydrogen bond*** (H bond). To be more precise, each water molecule can participate in ***four*** H bonds, as shown in Figure 16.3. As a first approximation we will regard these attractions as electrostatic ones, each between a lone pair and a proton, although the H bond apparently has sizable contributions from a variety of other factors — such as the covalent features described earlier.

Note the three-dimensional aspect: the five-water-molecule structure is tetrahedral, as can be seen by referring to Figure 14.9. Thus, liquid water and ice are vast three-dimensional

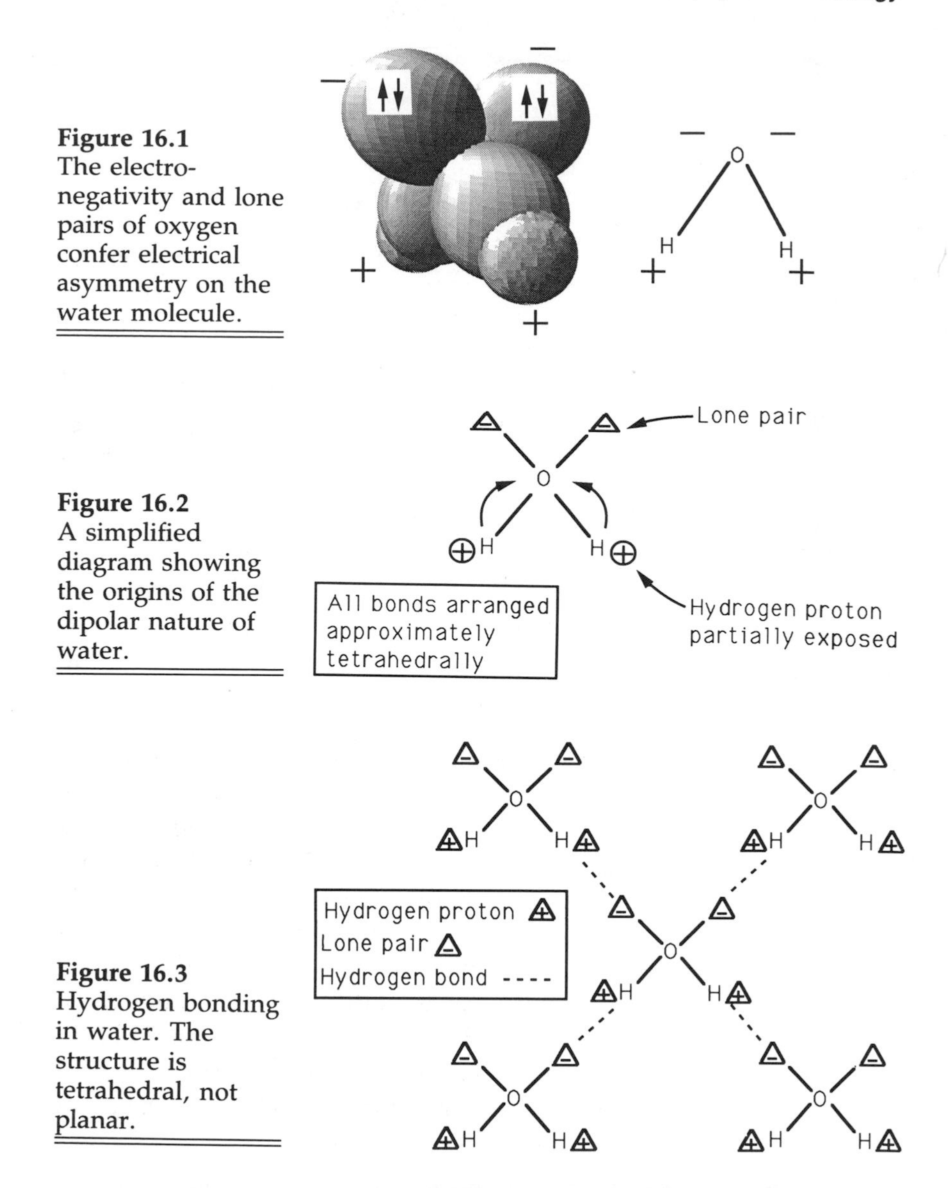

Figure 16.1 The electronegativity and lone pairs of oxygen confer electrical asymmetry on the water molecule.

Figure 16.2 A simplified diagram showing the origins of the dipolar nature of water.

Figure 16.3 Hydrogen bonding in water. The structure is tetrahedral, not planar.

lattices of molecules, each molecule being hydrogen-bonded to perhaps four others at any given time.

The energy needed to break an H bond is about that generated by the heat from an object at room temperature (30°C); H bonds are therefore of marginal stability at room temperature.

For example, water is a liquid at room temperature because some of its H bonds are constantly being made and broken (via heat exchange with their surroundings). H bonds are more stable below 0°C, resulting in the solid form, ice.

The Hydrogen Bonds of Water are Important in Biology

We can also now understand why water resists changes in temperature so well — heat energy which is added to water becomes partitioned out among water's many H bonds, either breaking them or changing their vibrational frequencies, thus diluting the effect of the heat. In reverse, removing heat from water does not change the water's temperature very much, compared to most other common compounds. The amount of heat required to change the temperature of 1 g of a substance by 1°C is the ***specific heat capacity***; for example, the heat capacity of water is 1 cal/(g C), meaning that 1 cal added to (or removed from) 1 g of water raises (lowers) the water's temperature by 1°C. For comparison, the heat capacities of most organic liquids, e.g., ethanol and glycerol, range from about 0.2 to 0.7 cal/(g C), meaning that 1 cal added to (removed from) 1 g of these compounds changes the temperature by 1.3 to 5.0°C. The heat capacities of most metals are less than about 0.1 cal/(g C).

Water's high heat capacity has far-reaching biological consequences. The water in a lake changes its temperature slowly when the air temperature above it goes up or down, thus protecting temperature-sensitive aquatic organisms. Large bodies of water and the nearby areas usually have smaller seasonal temperature fluctuations than do land-locked areas. The central U.S. has temperatures above 40°C in summer and below −30°C in winter; native organisms must be prepared for a long winter dormancy or hibernation. Much of the surface water is frozen and not available, even to evergreens, unless they have very deep roots. Along the coast, however, the temperature changes are more moderate; plants and animals often need only a short winter dormancy, or none at all.

In temperate areas the water in lakes may never freeze, even when the air temperature is below freezing for long periods. If the water does freeze, another physical property of water, also due to its many H bonds, exerts an effect: water has a high ***specific heat of fusion.*** Imagine that 1 g of water is at 3°C; now remove 3 cal to reduce the temperature to 0°C. The water will remain a liquid at 0°C until another 80 cal are removed. We say that the specific heat of fusion of water is 80 cal/g. At the other end of the scale, the ***specific heat of vaporization*** of water is 540 cal/g, meaning that 1 g of water at 99°C is heated to 100°C by the addition of 1 cal, but will remain a liquid at 100°C until ***another*** 540 cal are added. Thus, the evaporation of sweat exerts a powerful cooling effect by carrying away a considerable amount of body heat. Evaporative effects can cool an organism, even if the external temperature is higher than the organism's own temperature. If the external temperature is high and evaporation is restricted by high humidity or clothing, the organism could rapidly develop hyperthermia (there is a problem on this subject at the end of Chapter 27).

As water cools below 4°C, the H bonds extend somewhat and the water's density decreases. Thus, ice floats and the deep parts of a lake are the last to freeze, allowing certain aquatic organisms to survive the winter.

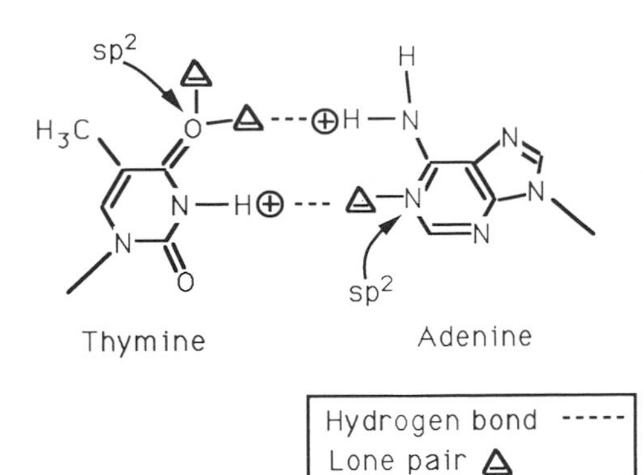

Figure 16.4
An example of hydrogen bonding in DNA.

A Variety of Atoms Can Participate in Hydrogen Bonding

An H bond potentially can form at any time there is a correct orientation between the two requisite groups — a lone electron pair and a hydrogen atom (the latter attached to a carbon, nitrogen, or oxygen, all of which are more electronegative than hydrogen). Thus, in biological systems opportunities for hydrogen bonding abound, so much so that the large number of H bonds which appear can often generate stable structures — in spite of the relative weakness of the individual H bonds. Figure 16.4 shows the hydrogen bonding between adenine and thymine in DNA; H bonds between lone pairs and hydrogen protons are labeled. The specificity of base pairing is explained by the position and polarities of the participating moieties, meaning that no H bond could form if two lone pairs or two hydrogen atoms faced each other. The only base pairs with the requisite positions and polarities are adenine-thymine (AT) and cytosine-guanine (GC).

A typical molecule of DNA is held in the double-helical configuration by millions of H bonds among the millions of nucleotide pairs, the strength of the latter being in their numbers

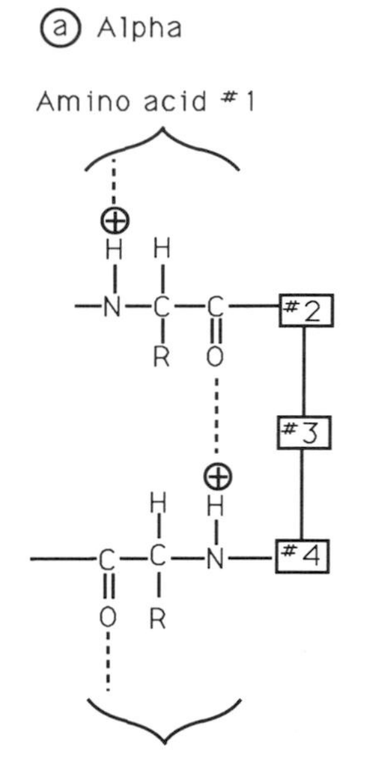

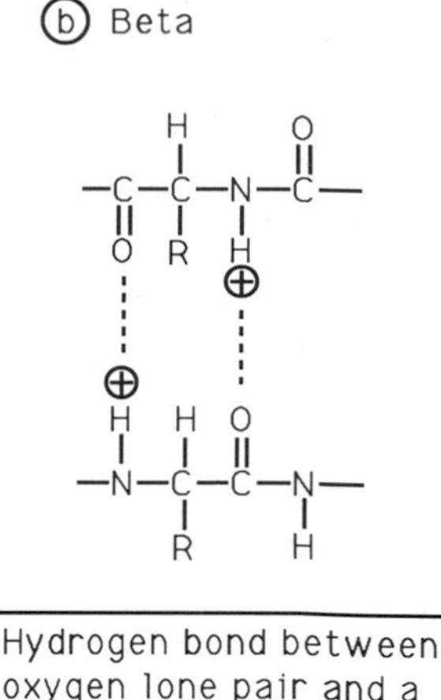

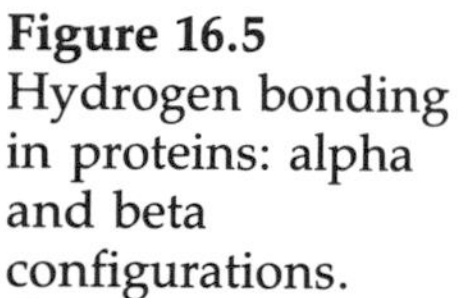

Figure 16.5 Hydrogen bonding in proteins: alpha and beta configurations.

rather than in their individual properties. We might expect, however, that enough heat energy would break sufficient H bonds to cause separation of the halves of the double helix (***denaturation***). This expectation is borne out by experiments in which DNA is shown to denature at 45 to 60°C. Of greater interest is that this marginal stability of DNA at biological temperatures (below 45°C) facilitates the separation of the two strands of DNA during DNA synthesis.

Figure 16.5a is a simplified picture of the hydrogen bonding between the hydrogen of a peptide linkage and the double-bonded oxygen of the residue four removed along an alpha helix. Each residue can participate in two such H bonds, meaning that a typical polypeptide may have several hundred. Figure 16.5b shows the H bonds of a beta-pleated sheet in a polypeptide; again, each residue participates in two H bonds. There are enough

H bonds in typical proteins to maintain the secondary and some of the tertiary structures of those complex compounds.

The great diversity of water-soluble compounds can be explained by water's hydrogen bonding capability; that subject is discussed in Chapter 19.

Applications, Further Discussion, and Additional Reading

1. An extensive, but nonmathematical, treatment of hydrogen bonding is contained in *Chemical Bonding Clarified Through Quantum Mechanics,* by Pimentel, G. C. and Spratley, R. D., Holden-Day, San Francisco, 1969.
2. A somewhat technical treatment of hydrogen bonding can be found in *Hydrogen Bonding,* by Josten, M. D. and Schaad, L. J., Marcel Dekker, New York, 1974. A second technical reference is *Hydrogen Bonding,* by Vinogradov, S. N. and Linnell, R. H., Van Nostrand Reinhold, New York, 1971.
3. When a person is immersed in a large body of cold water, say following an accident at sea, that person's body heat is transferred to the cold water and ***hypothermia*** results. The limited ability of the person's body to generate heat, the great volume of water and its high heat capacity prevent the water near the person from heating up. Thus, heat continues to flow from the person to the water, cooling the person. The person's chemical reactions, which are designed to run at 37°C, slow down and death can result very quickly. On the other hand, controlled hypothermia can be used medically to reduce a patient's need for oxygen during heart surgery.

4. Heat denaturation of DNA is the result of breaking the H bonds that hold the two DNA polymers in the double helical configuration. Adenine-thymine (AT) pairs share two H bonds and cytosine-guanine (CG) pairs share three H bonds. Thus, we would expect that CG-rich DNA would have a higher denaturation temperature than AT-rich DNA, an expectation that is borne out by experiments.

5. The denaturation of DNA can be followed accurately by observing its ultraviolet (UV) absorption properties. When two nucleotides are hydrogen-bonded they form a local region of very high concentration. The UV absorbing properties of nucleotides change at high concentration and return to "normal" upon denaturation. Thus a graph of UV absorption vs. temperature shows a dramatic change in absorbance of DNA at the denaturation temperature. See Chapter 18 for a discussion of light absorption by biochemicals.

6. As discussed in Chapter 14, the two nuclei at the ends of a bond seem to oscillate with respect to one another, and the energy of such oscillations is quantized. This is as true for H bonds as for covalent bonds, and the IR absorption spectra for H bonds has been intensively studied. Thus, there exist considerable data on the IR absorption of specific H bonds and also that of the covalent bonds adjacent to them. Changes in these spectra can yield information about the existence and the strength of the H bonds under varying physical conditions. More information is contained in the books cited in item 2, above.

Chapter 17
VAN DER WAALS' INTERACTIONS

SOME IMPORTANT BONDS ARE VERY WEAK

Consider the case of cyclohexane at room temperature: the molecules obviously attract one another because cyclohexane is a liquid. Yet it has no lone pairs and therefore no hydrogen bonds. What is the source of intermolecular attraction?

Cyclohexane and other hydrocarbons have virtually no permanent electrical charge asymmetry. Their electrons are in constant motion, however, and for extremely short time periods there may result slight charge asymmetries, as shown in Figure 17.1. These charge asymmetries can make one part of the hydrocarbon molecule positive and another part negative, i.e., electrical dipoles are created. The weak, short-lived dipoles are called ***transient*** dipoles and two of them can attract one another. That attraction is called a transient dipole-transient dipole attraction. In addition, one transient dipole can ***induce*** a transient dipole in another molecule and then be attracted to it.

These transient dipole-transient dipole and transient dipole-induced transient dipole attractions are collectively called ***van der Waals'*** interactions. Their transience makes them very

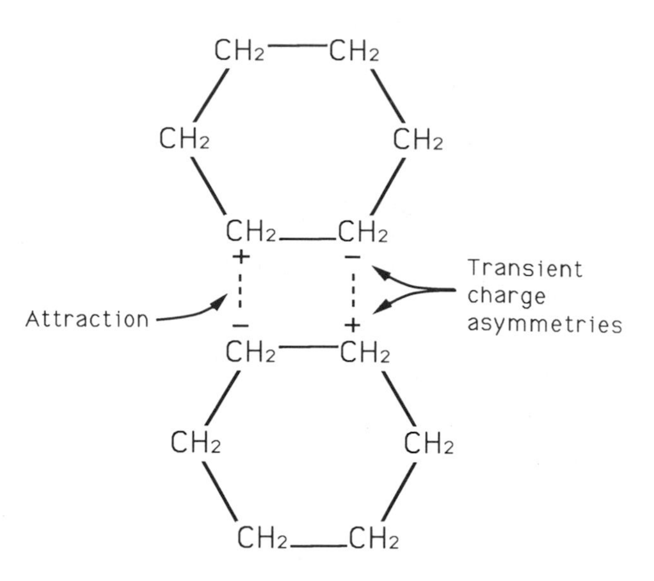

Figure 17.1 Transient interactions in cyclohexane. These charge asymmetries arise from the natural motions of charges in molecules. They are short lived, but common enough to provide enough attraction to hold many hydrocarbons together in solution.

weak — perhaps of the order of 10^{-2} eV — but in the absence of the strong attractions associated with permanent monopoles and dipoles, they are the only attractive forces available for keeping the hydrocarbon in the liquid phase — if indeed it *is* a liquid. Further, ***any*** compound can engage in van der Waals' interactions; thus their sheer numbers can cause them to be a serious concern in questions of intermolecular interactions.

Applications, Further Discussion, and Additional Reading

1. The nomenclature regarding nonspecific transient interactions is somewhat confused. It is not uncommon to find the name "van der Waal" used to refer to only a single kind of such interaction, rather than in the generic sense used here.

Chapter 18
The Absorption Spectrophotometer

Structure Determines the Energy of Light Absorption

In this chapter we will examine the absorption spectrophotometer, a device for the quantitative study of ***electronic transitions*** (although the same general methods are used for vibrational and rotational transitions). The aim is to apply and knit together the theoretical principles presented in the earlier chapters; thus, some algebra is involved.

There are two basic ideas behind the use of the absorption spectrophotometer. First, the specific light energy that a chemical absorbs can be used to identify that chemical and, second, the amount of light absorbed can be used to measure the concentration of the chemical. These two concepts are discussed below.

The first consideration is the actual energy absorbed. A molecule or atom will absorb light if, and only if, the energy of that light is the same as the energy of separation between an occupied orbital and an orbital which is not fully occupied. In the simple model of Figure 18.1 the indicated transition will take place only if a quantum of energy of 3.5 eV is absorbed by the absorber. The transition from orbital 1 to orbital 2 will then take place.

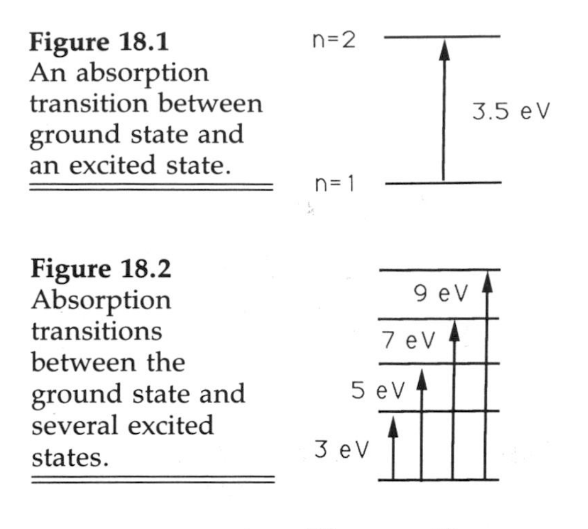

Figure 18.1
An absorption transition between ground state and an excited state.

Figure 18.2
Absorption transitions between the ground state and several excited states.

The specific energy of the transition will depend upon the structural details of the orbitals which, in turn, depend upon the identity of the atom or atoms involved. The energy may be determined, for instance, by the presence of carbon with a particular hybridization or by an oxygen atom in some specific configuration. Most molecules have several such transitions at several energies. For the hypothetical absorber in Figure 18.2, we would obtain transitions at 3, 5, 7, and 9 eV.

We see that the various absorption energies for a molecule should be specific for a given molecule's structure and that even a slight chemical modification to the molecule will change the energies absorbed. Thus, the specific energies absorbed are characteristic of the absorber and can therefore be used to identify the absorber. Figure 18.3 shows actual absorption spectra of two similar compounds — benzene and toluene. Their spectra are similar but not identical and could be used for identification purposes. Note that it is a property of molecules that their absorption spectra consist of wide "bands" of width perhaps 10 to 30 nm. Benzene and toluene show two such bands centered at about 210 and 250 nm. Very

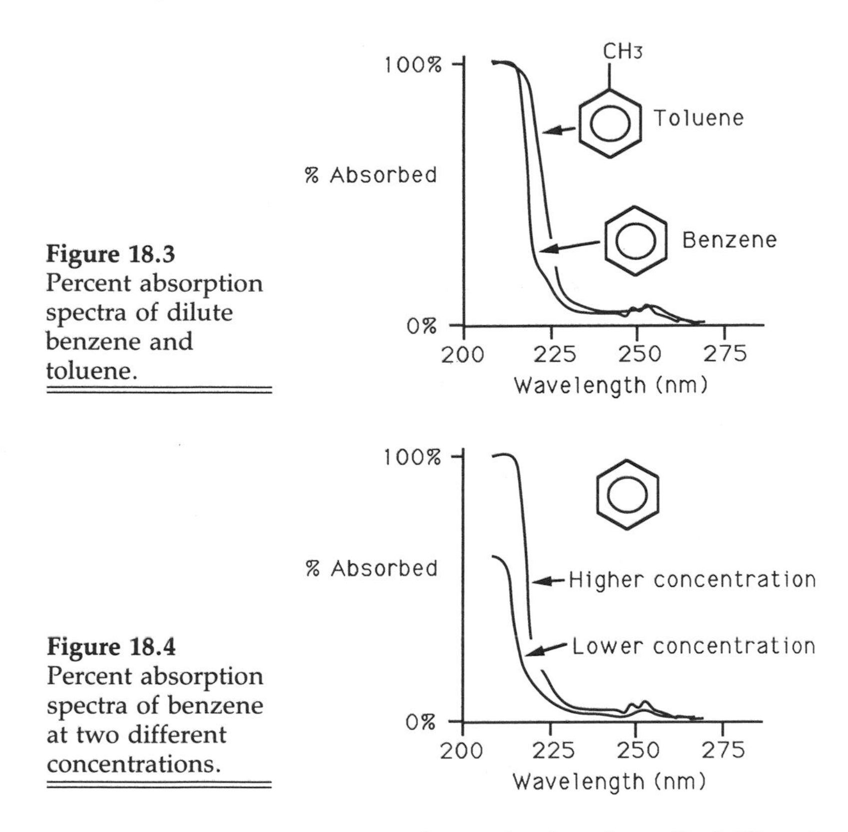

Figure 18.3
Percent absorption spectra of dilute benzene and toluene.

Figure 18.4
Percent absorption spectra of benzene at two different concentrations.

narrow absorption bands, called "lines", are characteristic of atoms.

INTENSITY OF ABSORPTION IS A MEASURE OF CONCENTRATION

The second consideration is the amount of light which is absorbed. If an absorber at a given concentration absorbs a given fraction of the incident radiation, then at a ***higher*** concentration ***more*** light should be absorbed because more absorbers are present in the light path. Figure 18.4 shows the percent of incident light absorbed by benzene at two different concentrations.

ENGINEERING DETAILS OF SPECTRO-PHOTOMETRY

In a schematic way an absorption spectrophotometer has the three parts shown in Figure 18.5. From left to right, the light source provides light of all the various energies needed to perform the measurement, the sample

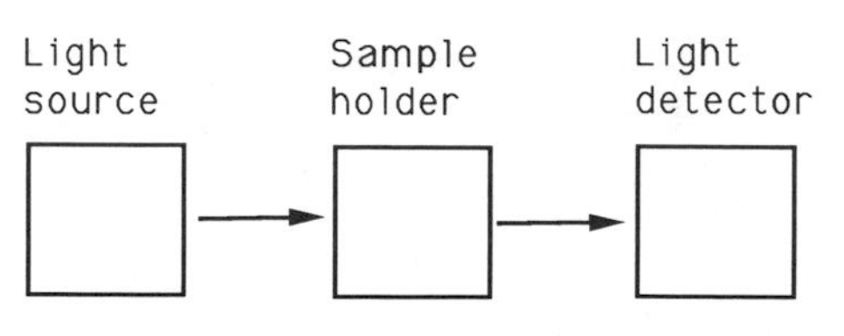

Figure 18.5 Simplified schematic diagram of a spectrophotometer.

holder is where the absorber is placed, and the light detector measures the amount of light that actually passes through the sample (i.e., is not absorbed). We now consider the three parts in more detail.

Light Source

This part must generate the various light energies which might be absorbed by the sample. The light energy, E, is related to the wavelength, λ, by

$$E = \frac{hc}{\lambda}$$

where h is Planck's constant and c is the speed of light. The numerator is just a constant:

$$E = \frac{1240}{\lambda}$$

where E is in electron volts and λ is in nanometers. Thus, a 4.0 eV quantum has a wavelength of (1240/4.0) = 310 nm. The reason for the conversion to wavelength is that, for mechanical reasons, it is easier to manipulate light spectra by wavelength than by energy, and we will therefore adopt that convention. Always keep in mind, however, that energy and wavelength are interconvertible.

A typical light source arrangement is shown in Figure 18.6. The incandescent bulb produces "white" light, which means it is a mixture of all the visible energies or, equivalently, a mixture of all the visible wavelengths. Our

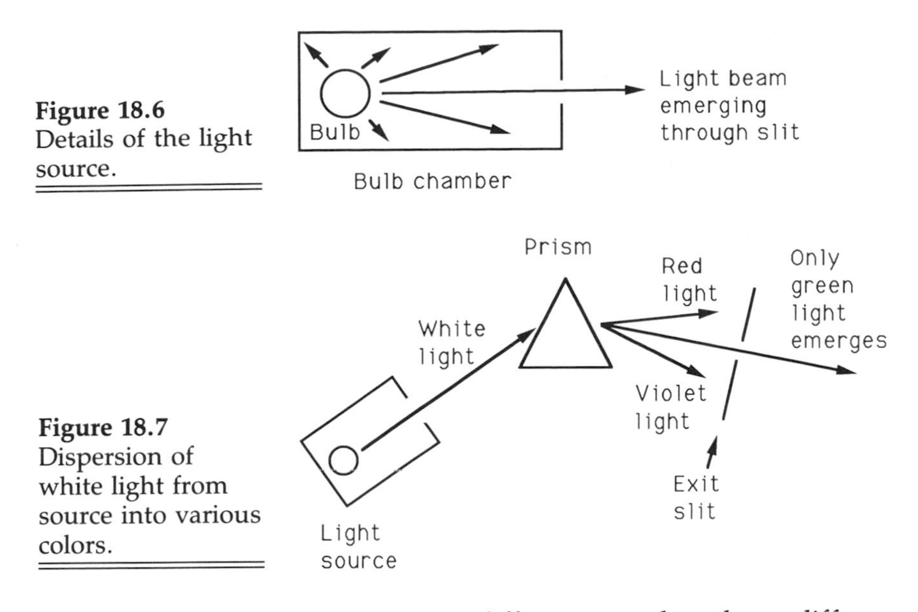

Figure 18.6
Details of the light source.

Figure 18.7
Dispersion of white light from source into various colors.

eyes perceive different wavelengths as different colors; blue is about 450 nm (2.75 eV), green is about 530 nm (2.34 eV), and red is about 625 nm (1.97 eV). Thus, "white" light can also be said to be a mixture of all the visible colors.

The white light from the bulb passes through a slit to get rid of any extraneous reflected light from the bulb chamber and then enters a prism. Prisms bend light according to the light's wavelength. This is demonstrated in Figure 18.7. Note that only green light ($\lambda \approx 530$ nm) emerges from the light source because of the placement of the exit slit. By rotating the prism clockwise, we can get red light instead of green light out of the second slit. This is demonstrated in Figure 18.8. You may have correctly guessed that some other parts also need to move, but the basic idea is to rotate the prism. Thus, we can select a particular wavelength (energy, color) that we want by rotating the prism. The light emerging from the second slit has a relatively limited range of wavelengths

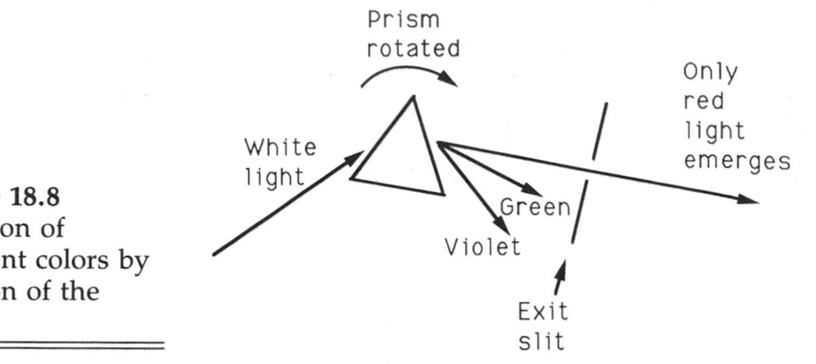

Figure 18.8 Selection of different colors by rotation of the prism.

and is called "monochromatic." A device that produces monochromatic light is a ***monochromator.***

Two modifications may be important in practice. First, the prism is often replaced by a mirror scribed with fine parallel lines. This device is called a ***grating,*** and it does what a prism does — disperses light into its component colors (wavelengths, energies). Because a grating ***reflects*** light from its front surface, it can be used for dispersing light that would be absorbed while passing ***through*** a typical glass prism, e.g., ultraviolet (λ < 320 nm). Absorption of incident light would mean that the prism would be worthless for dispersion.

Second, UV light is required for the spectrophotometry of most biochemical compounds — but UV is not produced by incandescent bulbs, and other types of bulbs must therefore be used. One example is a bulb that produces UV light by generating an electric field across mercury vapor. Mercury vapor bulbs have quartz envelopes because quartz does not absorb UV like glass does.

Sample Compartment

Usually this holds a glass (or quartz) container, called a cuvette, which has rectangular sides and a square horizontal cross section,

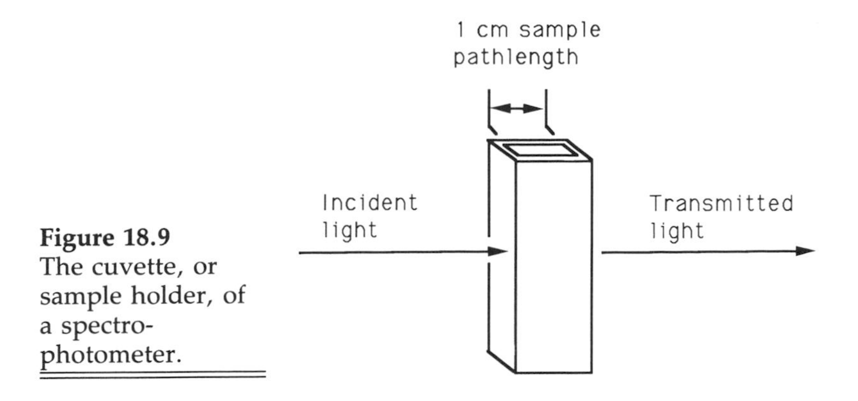

Figure 18.9
The cuvette, or sample holder, of a spectro-photometer.

allowing the incident monochromatic light to traverse 1 cm of path. The schematic arrangement is shown in Figure 18.9.

If the biochemical sample molecule does not have an electronic transition of the same energy as the light incident on it, all the light will emerge from the far (right) side of the cuvette. On the other hand, if the sample ***does*** have a transition of the same energy as the incident light, some of the incident light will be absorbed by the sample (to cause electronic transitions). Therefore, less light will emerge on the right than was incident on the left. We say that some of the light was absorbed and the rest was transmitted.

Light Detector

The amount of light transmitted through the sample is generally measured by a ***photomultiplier tube***. The light that strikes a photomultiplier generates an electrical current in the tube in proportion to the amount of light striking the tube. That current can then be measured by a current-measuring meter. The initial conversion of light energy to electrical energy takes place when a light quantum strikes a metal plate, called a photocathode, thus knocking an electron out of the photocathode. (Electrons are only loosely held by metals. That is why metals are good electrical

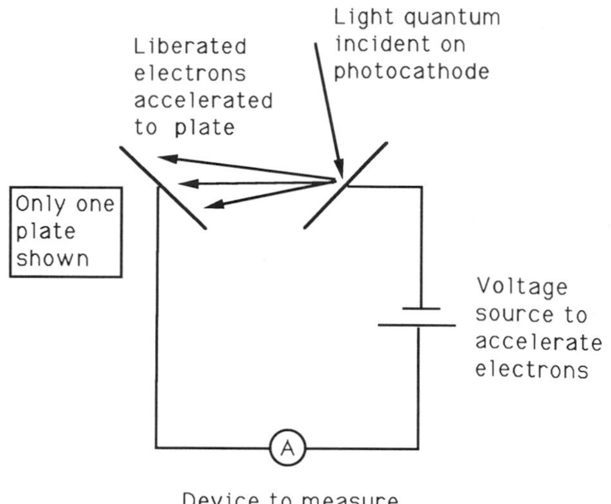

Figure 18.10
A simple photomultiplier, showing one electron liberating several electrons. An actual photomultiplier would have up to 14 such plates in series, each providing a multiplication factor of close to 10.

conductors and why light can easily displace electrons from them. Recall the Einstein photoelectric effect.) The electrons thus knocked loose are electrically accelerated to another plate to knock out even more electrons. Finally, the electrons are collected and the resultant current is measured. Photomultipliers may have as many as 14 consecutive accelerating plates, giving a gain of about 10^{12} (i.e., 10^{12} electrons out per quantum absorbed!). A photomultiplier is diagrammed in Figure 18.10.

An important property of photomultipliers is that their photocathodes tend to have specific ranges of sensitivity. A UV-sensitive tube, for example, may not be sensitive to red light and so it might be necessary to switch from a UV-sensitive photomultiplier to a red-sensitive photomultiplier part way through an experiment that requires both UV and red absorption measurements.

We can now "assemble" our spectrophotometer in more detail in Figure 18.11, as

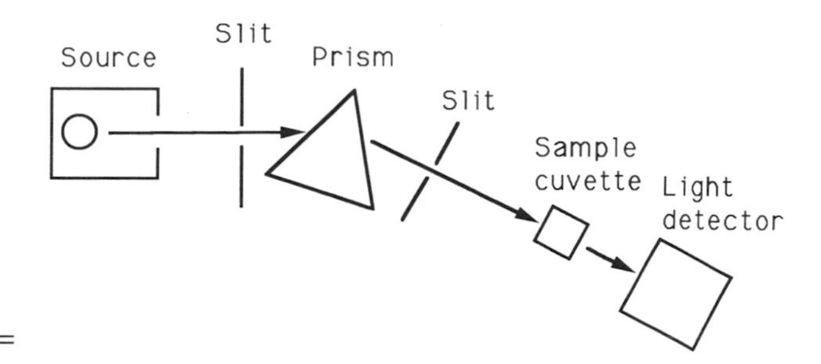

Figure 18.11 The assembled spectrophotometer.

previewed in Figure 18.5. If you examine an absorption spectrophotometer you can easily identify these parts, or their external presentations: there is usually a large dial marked in nanometers for wavelength selection — it rotates the grating; the sample holder is ordinarily under a small, light-tight trap door; and the intensity readout for the detector is a large meter face.

Quantitative Details of Spectrophotometry

We select a wavelength, place an empty sample holder between the light source and the detector, and then measure the intensity of light emerging from the source (incident light, I_i). We then put the actual sample into the holder and remeasure the intensity of light hitting the detector (transmitted light, I_t). The difference in the two measurements must be the light that was absorbed by the sample. The percent transmitted (%T) is

$$\%T = (I_t/I_i)100 \qquad (1)$$

This number ranges from 0 (if $I_t = 0$; all the incident light is absorbed) to 100 (if $I_i = I_t$; all the incident light is transmitted). The percent absorbed is $\%A = 100 - \%T$. Figure 18.12 shows a plot of %T and %A against λ for benzene. Evidently there is a weak electronic transition in benzene at λ = 250 nm (5.0 eV) and a very strong one at λ = 210 nm (5.9 eV).

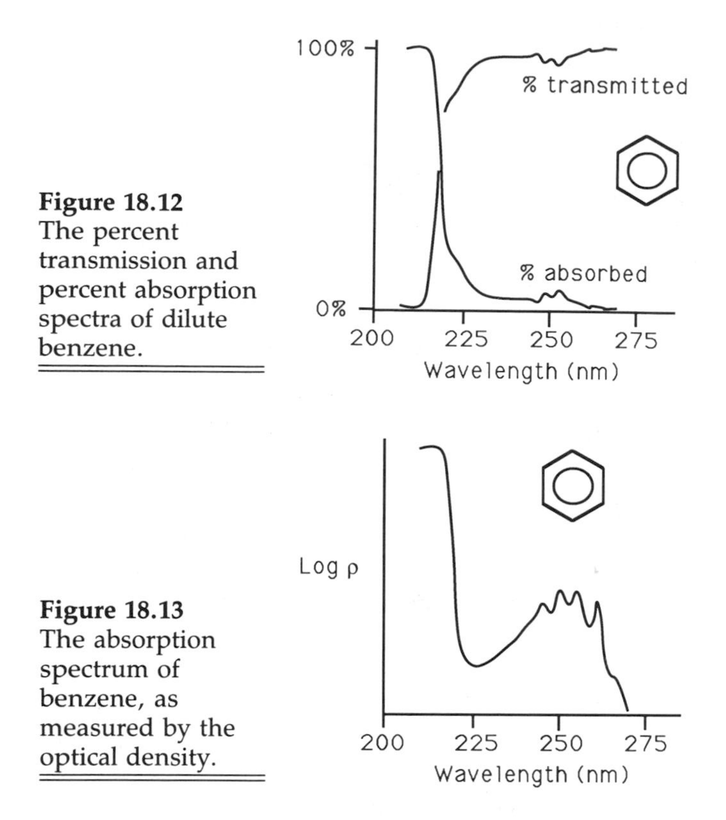

Figure 18.12 The percent transmission and percent absorption spectra of dilute benzene.

Figure 18.13 The absorption spectrum of benzene, as measured by the optical density.

Extensive experiments with light absorption show that, over a wide range of sample concentrations, %A = 100 − %T is ***not*** proportional to concentration. In other words, if you double the concentration of the sample, you increase %A but do not double it. In the case of liquids, with which one normally works, the usual measure of light absorbed is the ***optical density***, ρ, where ρ is related to the fraction transmitted by

$$I_t/I_i = 10^{-\rho} \qquad (2)$$

or equivalently, $\rho = -\log(I_t/I_i)$. It is ρ, rather than %A, which is proportional to the concentration. Figure 18.13 shows Log ρ plotted against λ for benzene. The data are merely that of Figure 18.12 converted with Equation 2.

Because absorber concentration is merely a measure of the number of absorber molecules (or atoms) in the light path, we can quickly see another way to increase the number of absorbers in the path — keep the concentration unchanged, but increase the path length through the solution. This yields

$$\rho = \epsilon c x$$

where c is the concentration, x is the path length through the sample, and ϵ is the proportionality constant that relates ρ to c and x. ϵ is variously called the absorption coefficient or ***molar extinction coefficient.*** The amount of light absorbed will depend on the wavelength used because, as noted earlier, whether or not light is absorbed depends on whether or not the light's energy corresponds to a possible electronic transition in the absorber molecule. This gives

$$\rho(\lambda) = \epsilon(\lambda) c x$$

Concentration and path length do not depend upon wavelength; the former are independent variables, completely controlled by the experimenter. Figure 18.14 is a further presentation of the data for benzene, this time plotting Log $\epsilon(\lambda)$ against λ.

Note that, if we were to double the concentration of benzene, the absorption spectrum of Figure 18.14 ***would not change*** because at any given wavelength

$$\epsilon = \frac{\rho}{cx}$$

Doubling c or x merely doubles ρ as well because ϵ depends only on wavelength and,

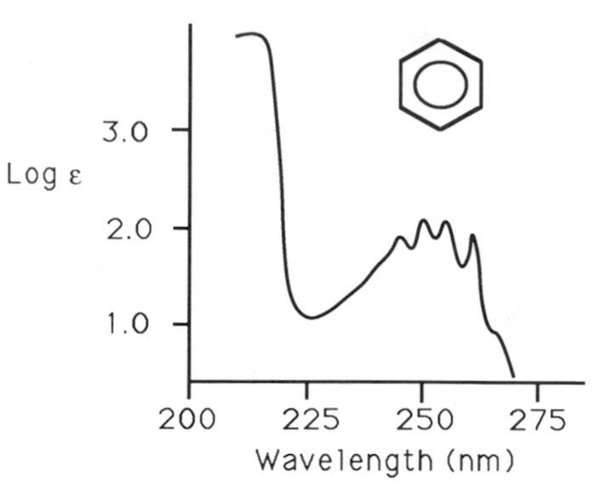

Figure 18.14 The absorption spectrum of benzene, as measured by the extinction coefficient.

therefore, remains unchanged. Because of this feature of ϵ, absorption data are almost always presented using the same axes as Figure 18.14, Log ϵ vs. λ. Further conventions are to use $x = 1$ cm(most cuvettes are of that size) and to use a value of c that keeps ρ (the optical density) less than about 1.0.

The Use of the Spectrophotometer to Measure Turbidity

The refractive index of a medium is a function of the electrical properties of the medium; in practical terms, a high index of refraction results in a lower velocity of light in the medium. When light enters a region of space in which there is a change of refractive index, the path of the light is bent. We see then that an optically inhomogeneous medium, which we say is "turbid", has a cloudy appearance because its inhomogenieties are bending incident light away from a straight line through the medium.

Refer to Figure 18.11; note that any light which starts out at the source but which does not reach the detector will be considered as light absorbed. Thus, any turbidity in the sample can ***scatter*** light out of the beam and into the light-absorbing surface of the sample compartment. No absorption by the sample would take place and yet a nonzero optical density would be read. Turbid samples should

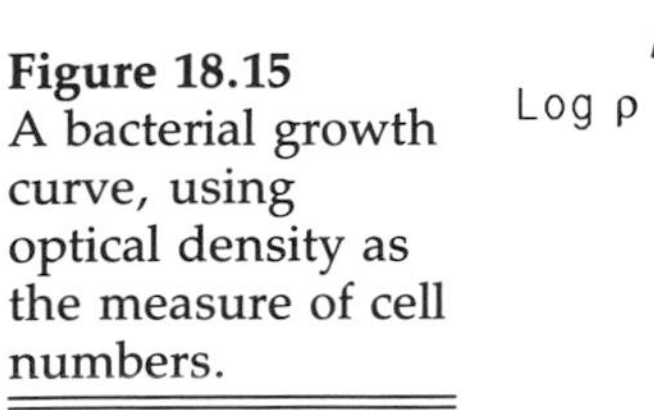

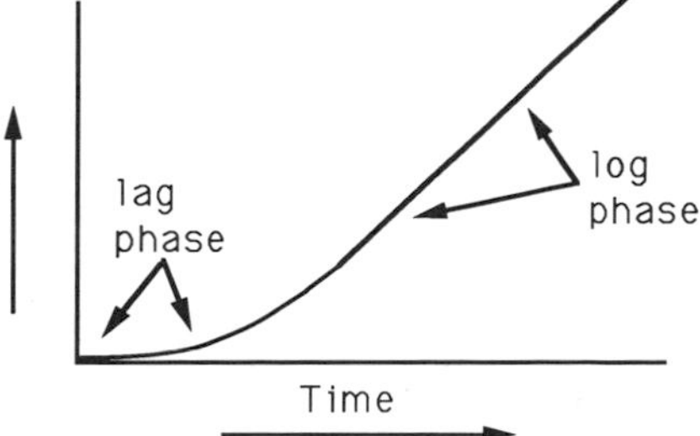

Figure 18.15
A bacterial growth curve, using optical density as the measure of cell numbers.

therefore normally be avoided. On the other hand, scattering is frequently constructively utilized in microbiology. A suspension of bacterial cells is turbid; it scatters light. The greater the concentration of bacteria, the more turbid will be the sample, and therefore the more light that will be scattered. Thus, bacterial concentrations can be measured on an absorbance spectrophotometer by observing the ***apparent*** optical density. One says "apparent" because very little actual absorption will take place; the spectrophotometer will be fooled by scattering and the use of the phrase "optical density" in such measurements is really a misnomer. That doesn't prevent the method from being useful, however. Figure 18.15 shows the "optical density" (really the turbidity) of a suspension of bacteria over a period of time. It is typical that there is an initial period of slow growth (called the "lag phase"), followed by a period of rapid growth (called the "log phase" because the logarithm of optical density plotted against the time is a straight line).

Applications, Further Discussion, and Additional Reading

1. A treatment of the principles of spectrophotometry is found in *Introduction to Research in Ultraviolet Photobiology,* by Jagger, J., Prentice-Hall, Englewood Cliffs, NJ, 1967. Another good source, also covering other aspects of photobiology, is *The Science of Photobiology,* Smith, K. C., Ed., Plenum Press, New York.

2. A biologist, using a 1-cm path length cuvette containing a 1-mM solution of a compound, measures the optical density to be 0.35. What is the molar extinction coefficient of the compound at that wavelength? (Answer: 350.) Next, the biologist is given a solution of the same compound and asked to calculate the concentration. The optical density of the new solution is measured to be 0.12. What is the new concentration? (Answer: 0.34 mM.) This is a simple, rapid, and accurate technique for measuring the concentration of a solution; applications are found in numerous biology laboratory exercises and in advanced work as well.

3. If a sample whose absorption is to be measured is dissolved in a solvent which interacts strongly with it, we should expect that the solvent will change the absorption properties of the sample. Thus, the absorption properties of the sample can depend upon factors other than its energy-level separations and its concentration. As an example, strongly acidic solvents frequently shift the absorption bands of samples by interacting with lone pair electrons in the sample. Bands which show such shifts must involve excitation of lone pairs into π electron clouds.

Chapter 19
SOLUBILITY

WATER-SOLUBLE COMPOUNDS FORM HYDROGEN BONDS TO WATER

The discussion in Chapter 16 makes it possible to understand the property of solubility. The presence of lone-pair electrons confers electrical asymmetry on electrically neutral molecules. Water provides a good example, in Figure 19.1 (compare Figure 16.2).

A given kind of molecule will be water soluble if that molecule's electrical asymmetry is complementary to the electrical asymmetry of water, thus providing a relatively stable interaction. Water is therefore "soluble in itself", as shown in Figure 19.2. The figure shown, consisting of five molecules, is planar in appearance, but, of course, must represent the actual tetrahedral form of water.

Using the same reasoning, we can see that methanol is highly soluble in water, as shown in Figure 19.3. As with water, the five-molecule structure is tetrahedral.

Sodium chloride (NaCl) is highly soluble in water. The stabilizing attraction between the sodium ion and the oxygen lone pair, and that between the chlorine ion and the hydrogen of water, are electrostatically a bit similar to hydrogen bonds, as shown in Figure 19.4.

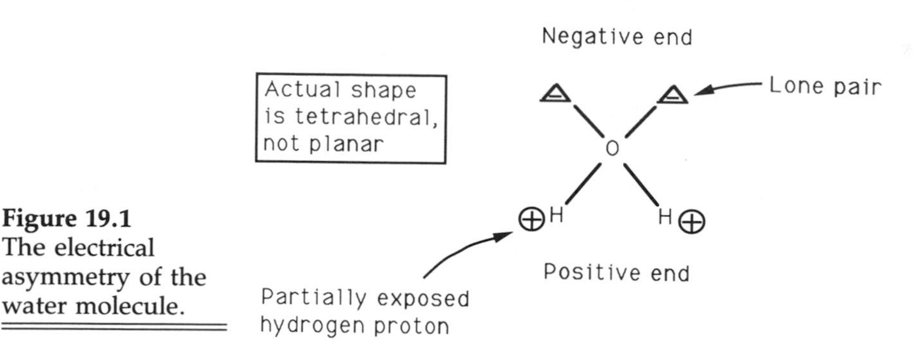

Figure 19.1 The electrical asymmetry of the water molecule.

The interaction between water and an ion is a ***dipole-monopole*** interaction. Water is so strongly attracted to the sodium and chloride ions (i.e., the monopoles) that those ions can seldom reassociate with each other in solution. Thus, the dissolution of compounds like NaCl in water is quite stable.

Hydrocarbons such as

$$H_3C{-}CH_2{-}CH_2{-}CH_2{-}CH_3$$

(*n*-pentane)

or

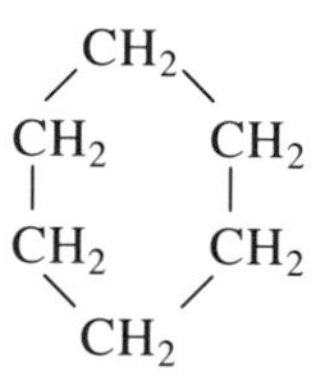

(cyclohexane)

have no lone pairs. There *is* some charge asymmetry in the carbon-hydrocarbon covalent bonds due to the electronegativity of the carbons; it is, however, insufficient in extent and orientation to allow these compounds to form strong H bonds to water. Thus, hydrocarbons are not water soluble to any significant degree. Hydrocarbons ***do*** interact with each other through van der Waals' interactions,

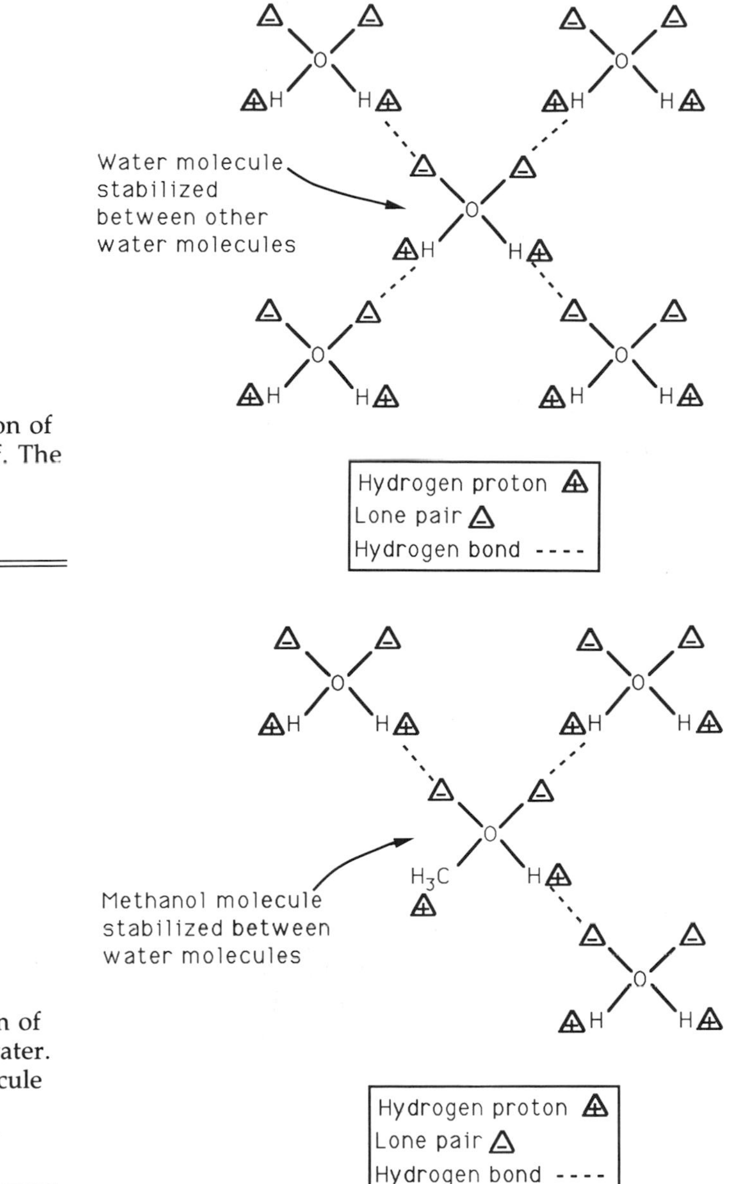

Figure 19.2
The dissolution of water in itself. The five-molecule structure is tetrahedral.

Figure 19.3
The dissolution of methanol in water. The four-molecule structure is approximately tetrahedral.

as described in Chapter 17, and benzene is highly soluble in cyclohexane, for example.

At a hydrocarbon-water interface a water molecule cannot form four hydrogen bonds to other water molecules because half of what

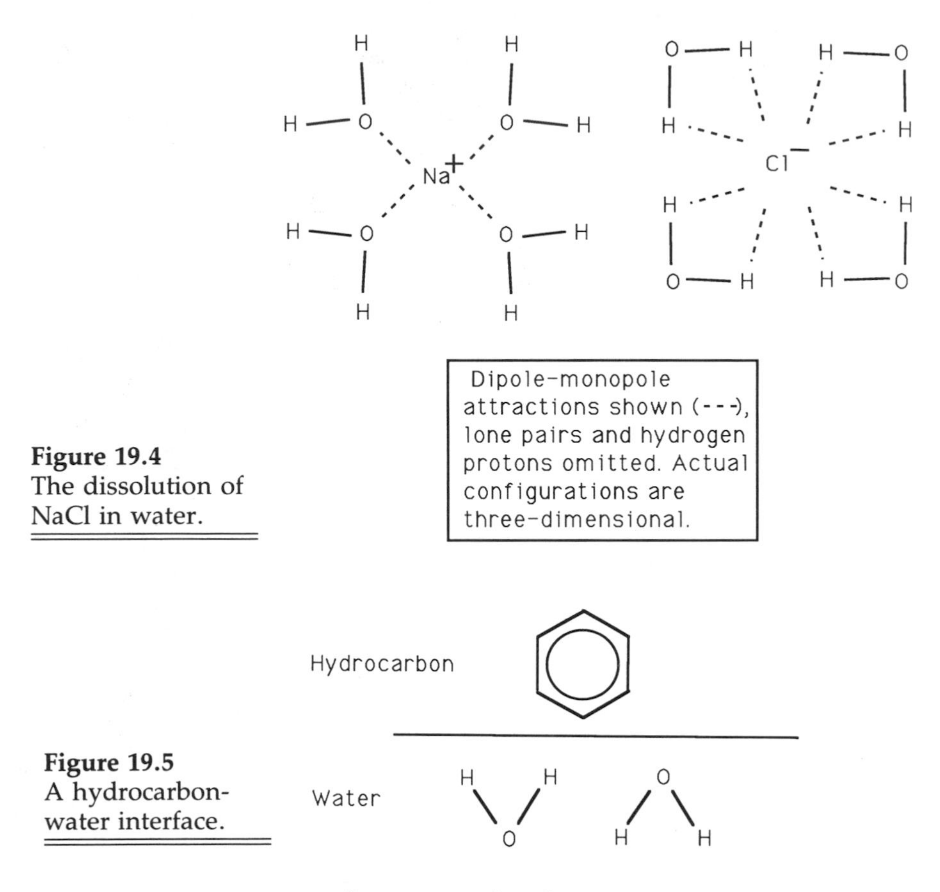

Figure 19.4
The dissolution of NaCl in water.

Figure 19.5
A hydrocarbon-water interface.

each water molecule "sees" at the interface is hydrocarbon (see Figure 19.5). This causes the water to form bonds "bent" into an energetically unfavorable shape that we might imagine to resemble that of Figure 19.6. The five-molecule tetrahedral configuration so typical of hydrogen-bonded water is distorted into other shapes to accommodate the hydrocarbon-water interface, where hydrogen bonds across the interface are not possible.

This energetically unfavorable distortion occurs only at the interface between the water and the nonpolar hydrocarbon. Thus, in order to minimize the number of distorted bonds, the smallest interfacial area between the water

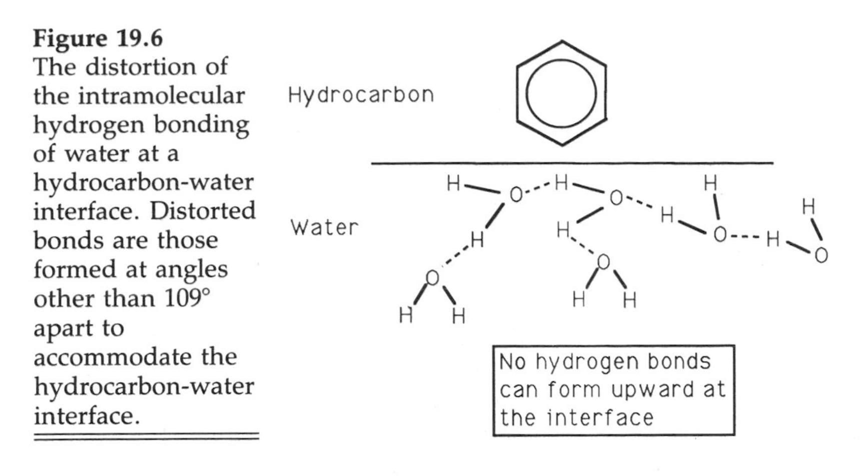

Figure 19.6
The distortion of the intramolecular hydrogen bonding of water at a hydrocarbon-water interface. Distorted bonds are those formed at angles other than 109° apart to accommodate the hydrocarbon-water interface.

and the hydrocarbon must be generated; this condition is met when the hydrocarbon forms a sphere. For example, salad oil consists of long-chain fatty acids; because of the long hydrocarbon tails on these acids, salad oil forms spherical globules in vinegar, the latter being mostly water.

Note that the spherical shape of the bulk hydrocarbon is forced onto it by the water's attempt to hydrogen bond to itself at the interface and is not the result of any new kind of bonding among the hydrocarbon molecules. Unfortunately, the phrase "hydrophobic bonds" is sometimes used to suggest that the spherical shape of the hydrocarbon clusters is held together by unique attractions. There ***are*** forces between the hydrocarbon molecules, as discussed in Chapter 17, but they do not account for the spherical clustering.

Applications, Further Discussion, and Additional Reading

1. A macroscopic analog of distorted chemical bonds: Figure 19.7a shows a weighted spring at rest vertically. That configuration is stable. Figure 19.7b shows the same spring and weight combination bent to the side. The configuration of Figure 19.7b is unstable, requiring the

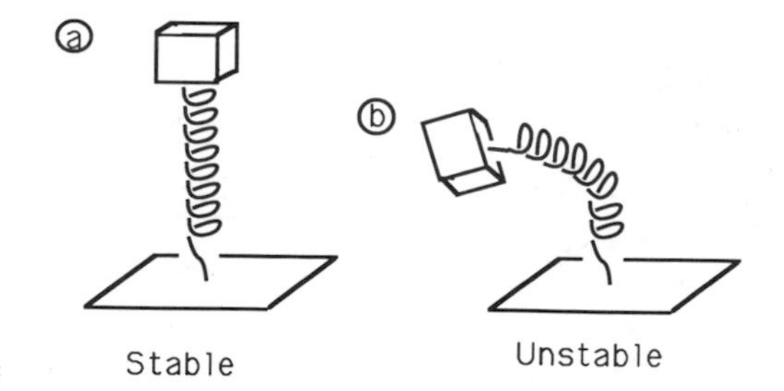

Figure 19.7
A stiff spring in a stable and an unstable configuration.

expenditure of energy for its maintenance. The spring in Figure 19.7a is an analog of a hydrogen bond in bulk water and the spring of Figure 19.7b is an analog of the hydrogen bond of water at a surface.

2. Alcohols have the structural formula R–OH. The OH group has two lone pairs, as shown in Figure 14.9, and this leads to great solubility of methanol in water, as shown in Figure 19.3. Imagine that the R group progressively becomes the ethyl, propyl, and longer hydrocarbon groups. The resultant solubility of the alcohol in water would be a compromise between the high solubility conferred by the OH group and the insolubility conferred by the hydrocarbon group. Thus, methanol and ethanol are infinitely soluble in water, but pentanol (five carbons) is virtually insoluble in water. These same considerations apply to the fatty acids, having the structure R–COOH.

3. Each of the 20 amino acids which occur in proteins has a side group that affects its solubility in water. In particular, those with a (nonpolar) hydrocarbon side group are essentially insoluble. Enzymes must function in the aqueous medium of a cell and therefore must themselves be water soluble. This would be impossible if the nonpolar amino acids appeared on the

outside of the tertiary structure of the enzyme. Thus, it is a common observation that most of the nonpolar amino acids of an enzyme are found in the interior of the molecule, where they can interact with each other via interactions between transient dipoles, as discussed in Chapter 17. The polar amino acids tend to be on the outside, where they can confer overall water solubility on the polypeptide. When a protein is heated, for instance, the heat energy disrupts the tertiary configuration, exposing the nonpolar residues to the aqueous surroundings; the protein then becomes insoluble, forming the familiar clot of fried egg white or boiled milk.

It was pointed out, in item 9 at the end of Chapter 14, that there is a free rotation about σ bonds, but that π bonds are fixed with respect to rotation. There are many σ bonds in a protein; the consequent freedom of rotation of the various parts makes, in principle at least, many tertiary configurations available to the macromolecule. The actual native, or functional, configuration that the protein ultimately adopts will be dictated by potential energy and/or entropic considerations. (See Chapters 20 and 21.)

Chapter 20
THERMODYNAMICS IN BIOLOGY

SYSTEMS, SURROUNDINGS, AND THE UNIVERSE

The literal meaning of "thermodynamics" is "heat power" and the subject usually covers several areas of interest to biologists, e.g., chemical equilibria and diffusion of molecules. We begin by defining some terms and then consider the two principles of thermodynamics.

A biologist may be interested in some ***local system***, e.g., a macromolecule, a test tube of material, a cell, or a deciduous forest. (The modifier, "local", merely indicates the somewhat limited extent of the system.) Each of these local systems is ***open*** because it can exchange matter and energy with its surroundings. Those surroundings consist of nonbiological and biological entities, the latter including the observer. The system and surroundings collectively constitute a ***universe***. A system which cannot exchange matter or energy with its surroundings is said to be ***isolated***. A universe is thus isolated (it has no surroundings) and a living system is open.

To be precise, the word "universe" ought to refer to the ***actual*** universe. This is impractically nebulous and we can safely limit the scope of our universe to the system plus those surroundings that have a perceptible effect on the local system in which we are interested. In this regard, we will usually assume that the surroundings are large enough to maintain constant pressure and temperature in the biological system of interest. In other words, the system will be isobaric and isothermal because the surroundings keep it that way. These conditions are typical of those encountered in biological research.

In terms of the local systems listed earlier, the practical surroundings might extend to a layer of water molecules around the macromolecule, an insulating wall around the test tube, a walk-in incubator to store cell cultures, or 2 mi^3 of space around a north Georgia forest. In any case, a well-thought-out description will always make clear the nature and extent of the local system and its practical surroundings, thus defining a practical universe.

THE PRINCIPLES OF THERMODYNAMICS

The ***First*** and ***Second Principles*** are strictly empirical; no exceptions are known and so they are often called "laws". They may be expressed in a variety of ways — the following are suitable for our needs.

The First Principle: Energy is Conserved in the Universe, But May Change Its Form

The conservation of energy is well known in physics, but we must be careful to include thermal energy in our energy accounting.

Suppose that we have an isolated system — no energy or matter can enter or leave; the energy of the system is called the ***internal energy***. The internal energy of that system can be changed by putting it in ***thermal contact*** with surroundings which have a higher temperature, causing heat to flow from the surroundings into the system. That heat flow increases the internal energy of the system. On the other hand, if the surroundings are colder than the system, heat will flow out of the system, thus decreasing the internal energy of the system.

The internal energy of the system can be changed in a second way: external work can be done on the system by the surroundings or by the system on the surroundings. As examples, external work may take the form of a change in the volume of the system or the addition or removal of material from the system.

The ***First Principle*** tells us that the energy of the universe remains constant, although that energy may move between the system and surroundings. There is no restriction made about the form the energy may take — merely that the total amount does not change. For instance, we could increase the internal energy of a system (a fireplace) by doing external work on it (e.g., add wood). Then we could turn around and decrease the internal energy of the system by removing heat (e.g., burn the wood). Overall the energy of the universe would have remained constant, the internal energy of the system would have returned to its original value, but heat energy and external work would have been interconverted.

We do not have to look far to find everyday examples of the conversion among energy forms, all of which are consistent with the ***First Principle***:

1. Mechanical to thermal — Rub your hand on a rough surface.
2. Electrical to thermal and optical — Turn on a light bulb.

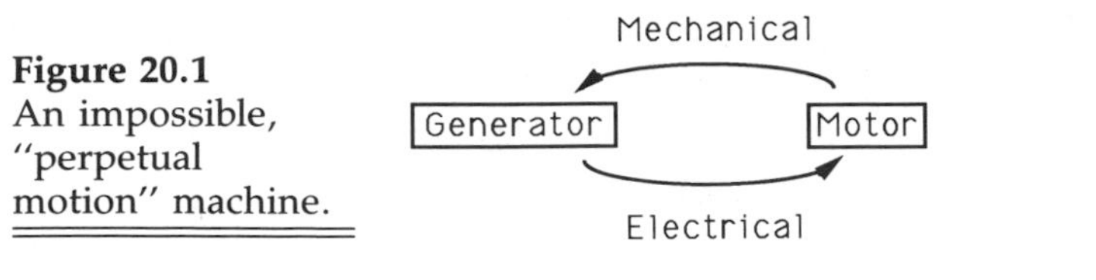

Figure 20.1
An impossible, "perpetual motion" machine.

3. Chemical to thermal and mechanical — Move muscles by running.
4. Thermal to mechanical — Expand a hot-air balloon.

The ***First Principle*** tells us about energy conservation, but makes no statement about another feature of energy, a feature we will call ***utility***. Utility refers to the ability of the (conserved) energy to do useful work, and it seems to be degraded during any process.

As an example, consider a generator driving a motor (electrical-to-mechanical conversion) and then the motor driving the generator (mechanical-to-electrical conversion), as shown in Figure 20.1. If that were the whole story we would have a perpetual motion machine! The problem, of course, is that the conservation of energy requires that we include the heat generated in each of these processes. Correctly worded, those processes are a generator driving a motor ("electrical-***plus heat***" to "mechanical-***plus-heat***" conversion) and a motor driving a generator ("mechanical-***plus-heat***" to "electrical-***plus-heat***" conversion). This is shown in Figure 20.2. Heat energy is irreversibly lost to the surroundings at each step, and while energy is surely conserved, the change to heat seems somehow to degrade some of the energy into a useless form, preventing the construction of a perpetual motion machine. In fact, heat generation is common to all four examples of energy conversion given in the previous paragraph, even in the fourth

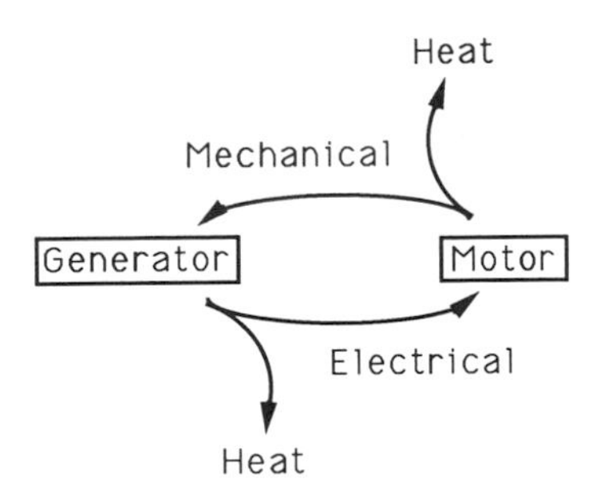

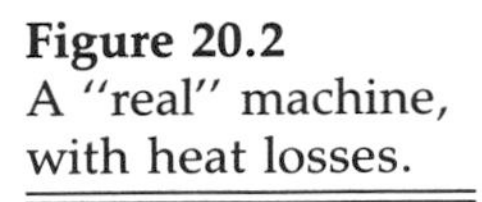
Figure 20.2
A "real" machine, with heat losses.

item (because the hot air in the balloon heats the surroundings).

It is not that heat energy per se is useless; the problem is that there is no way to recapture 100% of the heat lost to friction and then rechannel it back into, say, a steam generator to produce electricity. Even if we surround the system of Figure 20.2 with an insulator, so that the heat cannot escape, perpetual motion will not result. (See next paragraph.)

Other interesting examples of reduction in energy utility are easy to find. Suppose that a quantity of a gas is confined to a container with insulating walls and we suddenly open a valve leading into another identical container as shown in Figure 20.3. The gas will quickly distribute itself equally into the two containers. No heat can be exchanged with the surroundings because of the insulating walls and yet the process is one-way: the reverse process of spontaneous gas condensation into one of the two containers never occurs. Again we conclude that while energy has been conserved in the expansion the utility of that energy has been decreased — in this case without heat loss to the surroundings. (You ***will*** note that the temperature of the gas decreases if you have ever opened the valve of a compressed gas cylinder.)

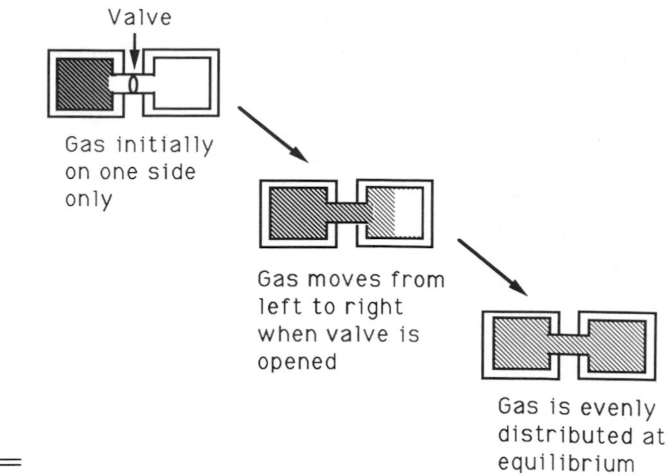

Figure 20.3
The expansion of an ideal gas into an insulated container.

The ***First Principle*** does not provide us with any information about the utility of energy, merely that the energy is conserved. The question of utility is addressed by the ***Second Principle***.

The Second Principle: Any Spontaneous Process Increases the Disorder of the Universe

A spontaneous process is one that takes place without any outside intervention. Examples of spontaneity are the rolling of an object downhill, the movement of heat from a hot body to a contacting cold body, and the movement of charge through a wire across battery terminals.

We next adopt the following working definition of ***disorder***: the most disordered of two configurations is the one which can be obtained in the most ways. As an example, suppose you had a board with 25 squares on it and could distribute four indistinguishable marbles, one to a square, in any way you chose just by dumping the marbles at random onto the board, as shown in Figure 20.4. Using probability theory we can calculate that there are 12,650 ways [(25) (24) (23) (22)/(4) (3) (2) (1)] to distribute the four indistinguishable objects among 25 positions.

Suppose now that we specifically restrict the marbles to only the four spaces in the upper right-hand corner of the board, as shown in Figure 20.5. This arrangement can be obtained in only one way (the one shown). By our working definition of disorder the system of Figure 20.4, tossing the marbles onto 25 squares, therefore generates more disorder than that of Figure 20.5. This conclusion agrees with our everyday notion of disorder: throwing the four marbles randomly onto a board of 25 squares makes a bigger mess (is more disorderly) than arranging the marbles neatly into one corner. Note that the configuration in Figure 20.5 is merely one of the 12,650 that would be obtained in the experiment of Figure 20.4.

Consider another ***specific*** configuration, as shown in Figure 20.6. This one specific configuration is just another one of the 12,650 calculated in considering Figure 20.4 and is exactly as probable as any other one configuration, ***including that of Figure 20.5***! In fact, ***every*** specific configuration is equally probable: the disorderliness generated in Figure 20.4 results from there being many possible outcomes, ***not*** to there being a high probability of a one specific "messy" outcome.

Consider a drop of ink at the edge of a container of water. This situation is analogous to that of Figure 20.5 (all the marbles at one corner of the board). Sometime later, the ink will be found to be distributed throughout the water, which is a situation analogous to the experiment of Figure 20.4 (marbles distributed helter skelter over the board). The distribution of the ink has changed from orderly to relatively disordered, as the ***Second Principle*** predicts.

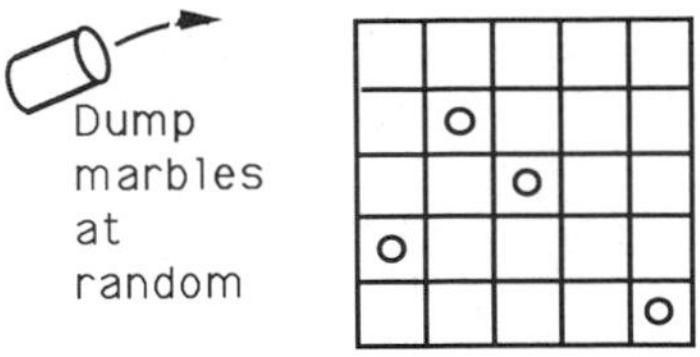

Figure 20.4
One random distribution of 4 marbles among 25 squares.

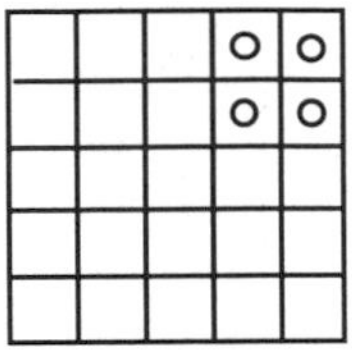

Figure 20.5
Four marbles are placed into four adjacent squares.

All this is not to say that the ink could not ***possibly*** revert back to a localized drop, but a localized drop represents only a few of the many configurations available, virtually all of which are "messy" in the sense of Figures 20.4 and 20.6, and all of which are equally probable. Thus, it is ***highly*** unlikely that the ink will ever find its way randomly back to a localized drop and in fact we never observe that to happen. These considerations now tell us why the gas which we allowed to expand into a larger container never retreated wholly back into the original container. It's not that it ***couldn't*** happen, but that such a retreat represents only a few of a huge number of equally likely configurations the gas could take, and therefore such a retreat is very improbable.

The diffusion of the ink and the expansion of the gas have a property in common, namely, that they are spontaneous, taking place without outside intervention. The reverse processes are not spontaneous, which of course is just a way of saying that they don't normally happen unless we somehow force them to happen. The ***Second Principle*** thus tells us that the universe is acting spontaneously in becoming more disorderly, i.e., in-

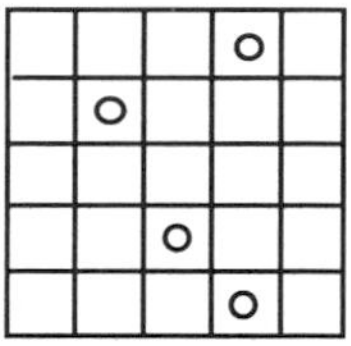

Figure 20.6 Another random distribution of 4 marbles among 25 squares.

			o	
	o			
		o		
			o	

creasing disorderliness is the natural direction of the natural evolution of the universe.

Physicists have a measure of disorder — they call it ***entropy***, and we can thus reword the ***Second Principle***: any spontaneous process increases the entropy of the universe. Some important ideas connected with the ***Second Principle*** are:

a. Entropy is not conserved the way energy is. Entropy is always increasing in the universe because spontaneous events are always occurring.

b. The ***Second Principle*** forbids the decrease of entropy in the universe, but entropy may decrease in some ***local system***, as long as there is a "more than compensating" entropy increase in the ***surroundings*** of that system. In that case, there will be a net entropy increase in the universe consisting of the system plus its surroundings. A green plant cell converts CO_2 gas (which has highly random molecular positions in space) into glucose (whose atoms are highly ordered into their positions by covalent bonds). Clearly the entropy of the atoms has decreased. This ordering, however, requires sunlight whose origin is in explosive reactions in the sun. Those nuclear reactions generate enormous amounts of entropy in the process, moving material in the sun into greatly randomized configurations.

If we add the entropy decrease in the atoms in the glucose to the entropy increase in the surroundings (which include the sun) there is a net increase, as predicted by the ***Second Principle***. In glycolysis and aerobic respiration, the energy liberated by glucose metabolism is used to bond inorganic phosphate to adenosine diphosphate (ADP) — decreasing entropy by covalently fixing atoms in place. At the same time, the entropy in the region surrounding the adenosine triphosphate (ATP) is increased by the conversion of glucose to CO_2 gas and by the metabolic heating of those surroundings, thus increasing molecular motion. In the universe comprising the mitochondrion and its surroundings the ***net*** change is an increase in entropy.

c. We have considered disorder as being a property of positional distribution, but there is another source of disorder — ***energy distribution***. The component atoms of glucose individually have little kinetic energy because their covalent bonds hold them in place. When the glucose is converted to CO_2 some of the covalent bond potential energy of the glucose is converted to kinetic energy, causing the CO_2 molecules to move and tumble about in space. Thus the energy, mostly localized to potential energy of bonds in glucose, becomes distributed among CO_2 bonds and into molecular kinetic energy of translation, vibration, and rotation. This redistribution of energy into more forms, or degrees of freedom, increases entropy just like the redistribution of the atoms in space does. The opposite occurs in photosynthesis.

d. In the discussion of the ***First Principle,*** we saw that, while energy is conserved, it seems somehow to be changed into a less useful form. Thus, after the expansion of the ideal gas into a larger isolated container the internal energy remained the same, but that energy could not push the gas back into the original container. The system of gas and containers was isolated and the ***Second Principle*** therefore tells us that the entropy of the gas was increased by the expansion, there being no surroundings. Evidently the loss of utility of the energy of the gas was connected with its entropy increase.

APPLICATIONS, FURTHER DISCUSSION, AND ADDITIONAL READING

1. The calculation of the number of arrangements of 4 indistinguishable objects among 25 spaces is explained in Chapter 4 of *Statistics for the Biological Sciences,* 2nd ed., by Scheffler, W. C., Addison-Wesley, Reading MA, 1979.

2. Suppose that there were a universe in which there were no entropy change. How would an observer in such a mythical world perceive the passage of time? A clock won't do because it is driven by the uncoiling of a spring or the discharge of a battery, both of which are irreversible, and winding or charging them will greatly increase the entropy. In fact, it has been suggested that our perception of time's passage depends on the relentless increase of entropy described by the ***Second Principle.*** You can find a discussion of this point in Stephen Hawking's book, *A Brief History of Time,* Bantam Books, Toronto, 1988.

3. The "classical" thermodynamic parameters — volume, pressure, and temperature — are defined at the macroscopic

level, and they really measure the ***average*** behavior of many different particles. We can see that such an approach ignores submicroscopic details if we consider a molecular gas: most of its volume is empty space; its pressure is due to collisions between the gas molecules and the walls of the enclosing container; and the temperature is a measure of the average kinetic energy of all the gas molecules that strike the thermometer. The field of ***statistical mechanics*** defines these same parameters, but remedies some of the deficiencies of the classical treatment by taking into specific account the fact that the molecules have a variety of energies and that the various molecules may interact with each other.

4. The thermonuclear reactions in the sun result in the fusion of hydrogen nuclei, forming helium. This kind of process obviously decreases entropy at the nucleus itself, but the amount of heat released is enormous and the overall entropy increases.

Chapter 21
THE FLOW OF ENERGY THROUGH A LIVING SYSTEM

EQUILIBRIUM STATES ARE THOSE OF HIGHEST ENTROPY

We will henceforth label the participants in a reaction by uppercase letters. The configuration, or ***state***, of a system will be given by listing the concentrations, or amounts, of all the participants at that time and will be denoted by a circled lower case letter. The conversion of one configuration to another will be called a ***process***.

We first look at an isolated mechanical system to get some insight into the nature of equilibria. Figure 21.1 shows the states of liquid in two containers. (Note that entropy increases ***downward***.) At the outset all the liquid is in the left container, and none is in the right container; this low entropy, or orderly, state is labeled ⓘ (initial). A valve between the two containers is then opened, the result of which is that some of the liquid spontaneously flows to the right — rapidly at first via an intermediate state ⓙ, then more slowly, until the liquid level is the same in the two containers, this latter state being one of higher entropy, ⓕ (final). The complete process is given by ⓘ → ⓙ → ⓕ, and the net process is given by ⓘ → ⓕ. Note again that the net process

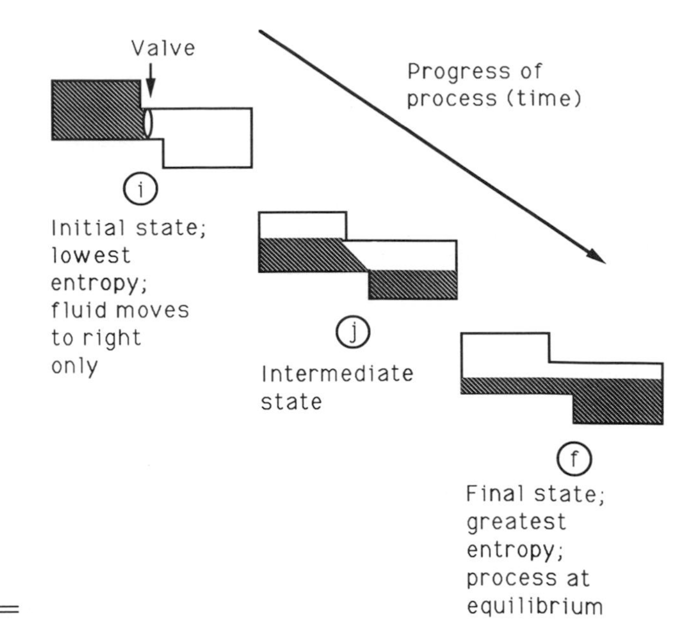

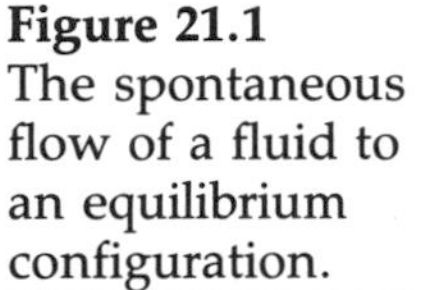

Figure 21.1
The spontaneous flow of a fluid to an equilibrium configuration.

is spontaneous, from low entropy state ⓘ to high entropy state ⓕ. Note also the resemblance to Figure 20.3

When the two liquid levels cease to change over time, and the system is free of outside intervention, we say that the flow is at ***equilibrium.*** (The nonintervention requirement will be described in the discussion of steady states, below.) Note that the equilibrium condition does not imply zero flow, only zero ***net*** flow, which leaves open the possibility that the two containers exchange material with each other at the same rates. The conversion of state ⓘ to state ⓕ is an ***irreversible*** process because the process ⓕ → ⓘ is never seen.

Suppose there is a biochemical reaction, A → B, where A is called the reactant(s) and B is called the product(s), and that the system is isolated. This is shown in Figure 21.2, which is a chemical analog of Figure 21.1. At the outset only A is present; thus, the reaction can initially go

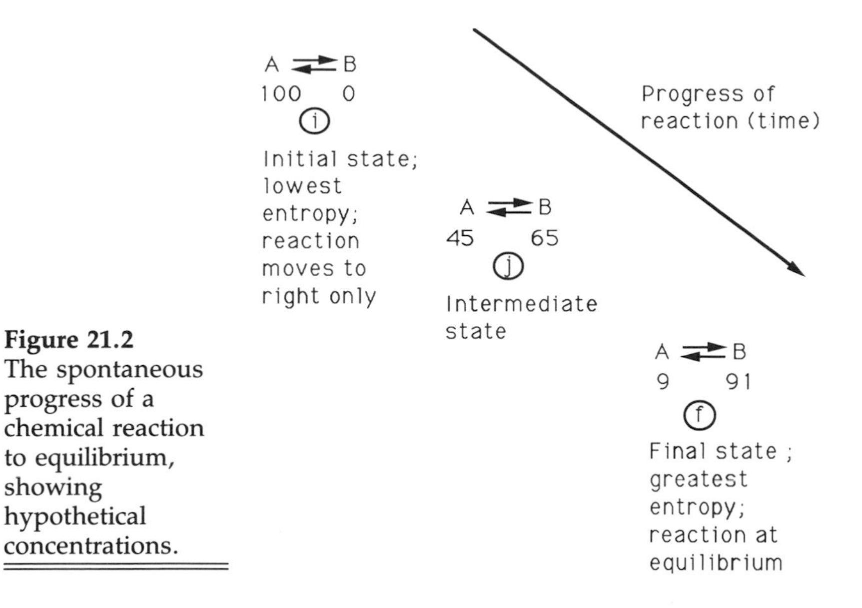

Figure 21.2 The spontaneous progress of a chemical reaction to equilibrium, showing hypothetical concentrations.

only in the direction of the formation of B. As B is formed the amount of B converted back to A will increase. Eventually the rate of interconversion of A and B will become equal and the ***relative*** amounts of A and B will cease to change; the reaction has reached equilibrium. Put another way, the process of conversion of state ⓘ to state ⓕ is spontaneous, or irreversible, and leads to an equilibrium mixture of participants A and B. (The arrows connecting A and B show that the reaction is ***potentially*** reversible, not that it is actually progressing in both directions. In state ⓘ the reaction can obviously go to the right only.)

Whatever equilibrium ratio of products to reactants is finally reached, we can be certain that ratio represents the situation of maximum entropy for the universe of the reaction. Otherwise, the ***Second Principle*** assures us that this universe would ***continue*** changing, generating entropy until it could produce no more. We thus interpret equilibrium (state ⓕ) as the state of maximum entropy.

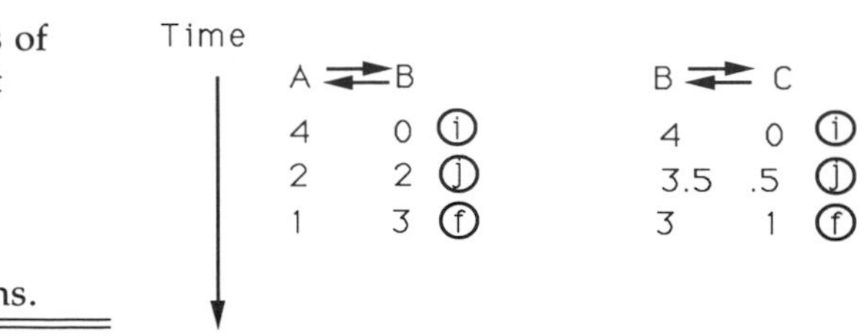

Figure 21.3 The progress of two different reactions to equilibrium, showing hypothetical concentrations.

Suppose, for illustration, that at equilibrium there is ten times as much B as there is A. The ratio of the concentrations of B to A is written as 10/1 = [B]/[A], where the numerator is the product concentration and the denominator is the reactant concentration. This ratio is called the ***equilibrium constant***.

A question should suggest itself: if B is more probable than A, why should ***any*** A remain at equilibrium? The answer lies in the ***Second Principle***. We started with A only, so that at the outset the reaction could go only to the right. However, as B was formed the concentration of A decreased, which gave the remaining molecules of A more space in which to distribute themselves, thus increasing the entropy and therefore the probability of A. The equilibrium state then became a compromise between the drive to create B and the drive to create A.

Consider now the two hypothetical reactions A ⇄ B and B ⇄ C, whose equilibrium constants are [B]/[A] = 3/1 and [C]/[B] = 1/3. Figure 21.3 shows the progress of these reactions when taken as separate systems.

We next take the same two reactions to be part of a single system in which B is a common intermediate, e.g., A ⇄ B ⇄ C. Figure 21.4 shows a simplified path toward equilibrium between the three participants. In state (i), only participant A is present; we then imagine

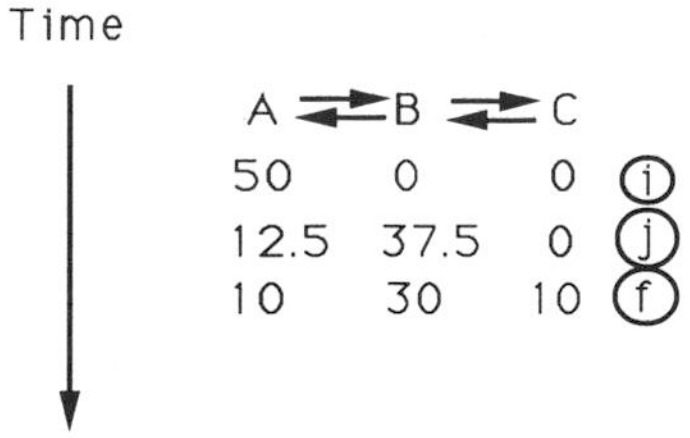

Figure 21.4
The progress of two sequential reactions, with the removal of product, leading to equilibrium.

Time ↓	A ⇄ B ⇄ C			
	50	0	0	ⓘ
	12.5	37.5	0	ⓙ
	10	30	10	ⓕ

that A ⇄ B first reaches equilibrium in state ⓙ, and that all participants then reach equilibrium in state ⓕ, where the concentrations of A, B, and C maximize the entropy of the system.

LIVING SYSTEMS ARE NOT AT EQUILIBRIUM

From Figure 21.4 we can see that the indicated reactions deplete A and create C, at least until equilibrium is reached. Assuming that A is a precursor and that C has some biological importance, the reaction sequence is thus biologically useful. However, things cannot just stop at that; after all, there is a ***continuous*** turnover of biologically important molecules in any living system. This instability manifests itself in many ways, some of which are

1. Organized biological structures will decompose at death, generating a variety of gases (for the most part).
2. Even in living systems, organized structures such as macromolecules are frequently unstable. For example, messenger RNAs and proteins have finite lifetimes, allowing the cell to shut down unnecessary reactions; DNA is denatured at every cell division; physical injury damages biopolymers; normal development requires the breakdown of biological order, e.g., the separation of fetal fingers.

For these reasons we expect that living systems will make frequent biological and

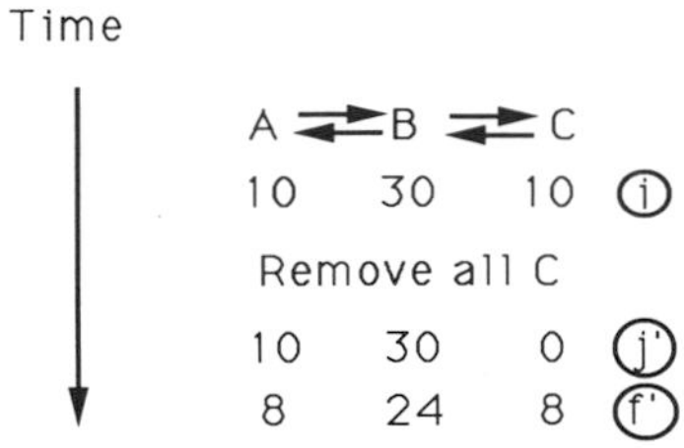

Time

A ⇄ B ⇄ C
10 30 10 (i)
Remove all C
10 30 0 (j')
8 24 8 (f')

Figure 21.5
The progress of two sequential reactions, with the removal of product, leading to equilibrium.

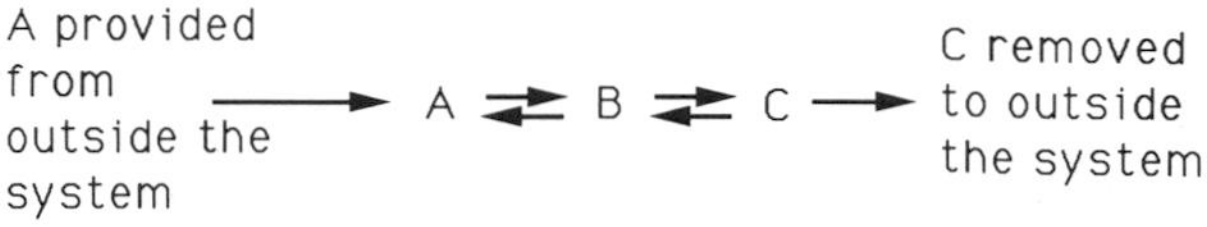

A provided from outside the system ⟶ A ⇄ B ⇄ C ⟶ C removed to outside the system

Figure 21.6
An open, steady-state system. Reactant is provided by the surroundings and product is removed to the surroundings.

physical demands on the concentrations of biologically useful molecules, such as C in Figure 21.4. To understand how the instability of C affects the system of Figure 21.4, let us irreversibly remove all ten units of C from stable state (f). (For example, we could imagine that C changes to an insoluble, therefore unreactive, compound.) The entire scheme is shown in Figure 21.5: the new intermediate state is (j'), which will be followed by a new equilibrium state (f'). Actually, (j) and (j') are identical to each other, as are (f) and (f'), the primed notation merely indicating that a fresh start must be made toward equilibrium after C is removed.

The new equilibrium state (f') has the same ratio of concentrations as did state (f), but the amounts of the participants have been reduced. There is thus a limit to the removal of C unless a source of A is provided. This leads us to Figure 21.6, which shows substance A being constantly provided from outside the system and product C constantly being removed from the system. We note that this system is thermodynamically open.

The concentrations of A, B, and C remain constant, but there is no equilibrium because the system is always being fed A while C is being removed. The system is said to be in a ***steady state*** and it forms a paradigm for all living systems, namely, a set of coupled reactions with constant input of reactants and constant removal of products. Compounds A, B, and C can ***never*** actually reach equilibrium in a living system: the state ⓕ cannot be actually attained because molecule A is added too rapidly and molecule C is removed too quickly. Thus, while the steady state concentrations of A, B, and C remain constant, those constant concentrations are not the same as those found at equilibrium; in the case of A it will be more and in the case of C it will be less. "Not-at-equilibrium" concentrations are typical of living systems and are maintained by the constant input of energy, e.g., reactants and sunlight, and the constant removal of products and heat.

If the organism represented by this system should die, the input of A and the removal of C would cease. At that point A, B, and C would become equilibrated, a condition characteristic of nonliving systems. In passing, we can now see why lack of outside intervention was a requisite for equilibrium when the latter was defined; the input of A and the output of C correspond to precisely such intervention.

Living Systems and the Flow of Orderliness

Having established the open nature of biological systems, we are now in a good position to understand a unique property they possess, namely, how they can maintain their low entropy in a universe whose entropy is always

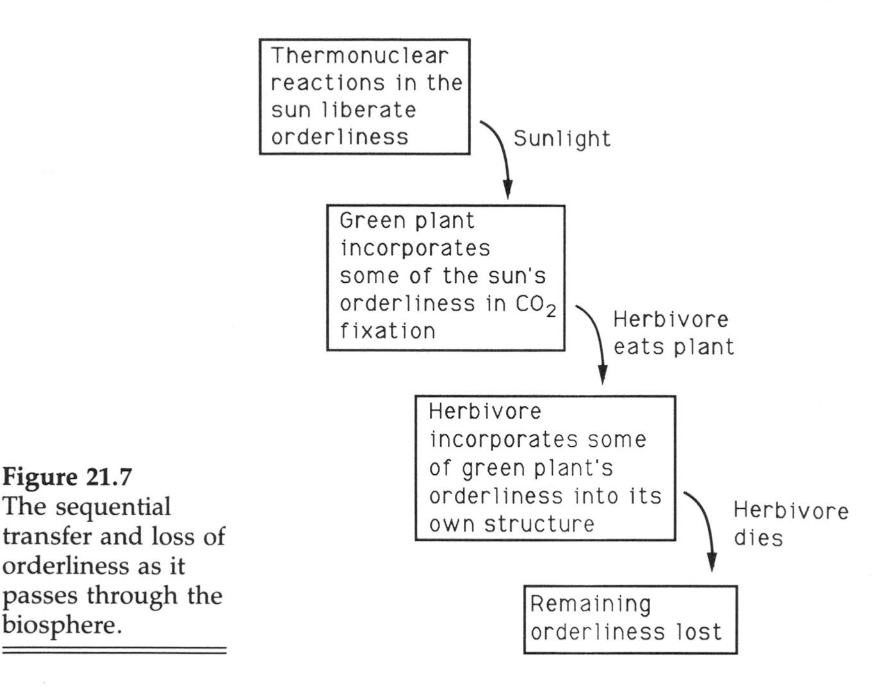

Figure 21.7
The sequential transfer and loss of orderliness as it passes through the biosphere.

increasing. First, as a convenience, let us adopt the word "orderliness" as a synonym for "low entropy". You should think of orderliness as a measureable quantity (a decrease in entropy is just an increase in orderliness). We next show that living systems, because they are open and interact with their surroundings, can be considered to input orderliness for their own purposes. Eventually that orderliness will be degraded and the organisms will give off disorderliness.

Figure 21.7 schematically shows a sequence of processes beginning with thermonuclear reactions in the sun and ending with the death of a herbivore (although any other process that generates local order would do as well).

We see here that the sun drives photosynthesis in a green plant on the earth. In the process, the sun's orderliness decreases while CO_2 is arranged into glucose, i.e., some of the orderliness originally in the sun has now become resident in glucose. Of course, photosynthesis is not 100% efficient and therefore some of the sun's orderliness is also lost to entropy production in the process.

Continuing, the green plant is eaten by a herbivore and oxidatively metabolized, converting glucose back to CO_2. This drives synthetic reactions in the herbivore, generating molecular order, which itself is eventually lost at death, via decomposition. All the original orderliness will ultimately be lost and heat will remain, generating disorder.

We could have skipped the herbivore steps in Figure 21.7 and allowed the plant to undergo oxidative metabolism, or we could have allowed a carnivore to eat the herbivore; no important consideration would have changed. Orderliness is passed from the sun to the green plant and ultimately some ends up temporarily in the biochemical molecules of a plant or animal; eventually all is lost to heat. Thus, living systems are open, taking in orderliness and discarding entropy.

ORGANISMAL PHYSIOLOGY AND "STANDARD" CONDITIONS

Refer to Figure 21.7, a schematic representation of the flow of orderliness from the sun through a simple food chain. Most physiology textbooks point out that about 38% of the solar energy absorbed by a plant winds up in glucose ***under standard conditions***.

"Standard conditions" means particular, fixed concentrations of reactants and products, specific temperature, and specific pH. For example, an athletic analog of conversion to standard conditions would be the time a person would need to run a mile if the temperature were 70°F, if there were no tail wind, and if there were a neutral crowd — instead of the time of the run under the actual, existing conditions of the track meet.

The notion of standard conditions is a very useful one, allowing experimental data to be expressed in common terms under which all data can be fairly compared. The problem here is that actual conditions in a cell are not known to be "standard". Even within a single cell various allied reactions can take place in different environmental "compartments", such as those maintained by the endoplasmic reticulum. We would be hard pressed to determine the actual experimental conditions in these microscopic compartments.

Returning now to Figure 21.7 and the quoted 38% efficiency of photosynthesis under standard conditions, it has been estimated by Nobel that the ***actual*** efficiency of conversion of absorbed solar energy into stored energy by plants on the earth is about 1%! Thus, at the top of Figure 21.7 we could expect that about 99% of the solar energy absorbed by plants will be lost to entropy.

Applications, Further Discussion, and Additional Reading

1. There is a good discussion of the efficiency of photosynthesis, under standard and field conditions, in Chapter 6 of *Biophysical Plant Physiology and Ecology,* by Nobel, P. S., W. H. Freeman, San Francisco, 1983.
2. The relationship of entropy to biology is covered in detail in two books by Harold J. Morowitz, *Energy Flow in Biology,* Academic Press, New York, 1968, and *Entropy for Biologists,* Academic Press, New York, 1970. These books include discussions of both traditional thermodynamics and of statistical mechanics.

3. We viewed a steady-state system as one with a steady flow of energy and/or material in and a corresponding flow out. In the limiting case where the flow is zero, the system would reach equilibrium. Thus, an equilibrium state can be thought of as a steady state of zero flow in and out.

4. There are many examples of everyday systems which are not at equilibrium. For example, an ordinary pendulum held off-center is kept away from equilibrium by the constant input of energy from someone's hand. A cloth windsock is at equilibrium when it hangs vertically; it extends horizontally only as long as it is kept there by energy from the wind.

5. The concept of the "flow" of orderliness through biological systems was first put into print by the famous physicist, E. Schrodinger, in his book entitled *What is Life?* (Cambridge University Press, London, 1945). Morowitz has restated this concept by pointing out that living organisms, being low entropy systems, are improbable and that this improbable condition is maintained by a constant inflow of energy from the sun to a sink. The price of the low entropy of an organism is an increase in entropy in the rest of the universe (Morowitz, H. J., *Energy Flow In Biology*, Academic Press, New York, 1968, p. 19). You can read more about Schrodinger's idea and its detractors in the articles by Pauling and Perutz in *Schrodinger: Centenary Celebration of a Polymath*, by Kilmister, C. W., Ed., Cambridge University Press, Cambridge, 1987, and in Thinking of biology: *What is Life?* revisited, by Sarkar, S., *BioScience*, 41(9), 631–634, 1991.

6. Suppose a herbivore is eaten by a carnivore and that carnivore is eaten by another carnivore, etc. By the time the final consumer's molecules have been polymerized, the original orderliness from the sun will have been reduced to a tiny fraction of its original value. Thus, a large quantity of plant orderliness is required to support even a small amount of orderliness in the second or third carnivore down the food chain. As one example of this principle, ecosystems generally support fewer predators than they do herbivores, although a few rapidly reproducing plants or herbivores may support a larger mass of predators.

A low entropy diet can be very expensive if it is obtained metabolically far from the sun in Figure 21.7. A much larger fraction of the orderliness generated by the sun can be obtained by consuming the plant than by consuming an animal that ate the plant. Vegetables and fruit are thus cheaper sources of orderliness than equivalent amounts of meat. In keeping with this, the wealth of a society generally correlates positively with the amount of meat eaten by its inhabitants. (It is ironic that high-meat diets also correlate positively with several serious health problems, e.g., cancer and cardiovascular disease.)

Chapter 22
FREE ENERGY

FREE ENERGY IS A PROPERTY OF THE SYSTEM ONLY

The concepts of energy conservation and entropy change are sufficient to explain the biological, biochemical, and physical phenomena in which we are interested. The quantitative use of the ***Second Principle***, however, necessitates that one keep tabs on the entropy of both the system and the surroundings. The latter quantity — entropy of the surroundings — is hard to work with because of the difficulty of defining a practical universe. Physical chemists circumvent this problem by using a quantity called ***free energy***. Free energy has the virtue that it can be described in terms of system parameters only, e.g., the temperature, volume, and entropy of the system.

We begin by specifying that we will use only the form of free energy proposed by Gibbs (it is not the only one). The practical use of Gibbs' free energy requires that the temperature and pressure of the system remain constant. These are mathematical requirements, but they need not bother us because they are also biologically realistic. If we add another biologically reasonable assumption — that the volume remains constant — ***the change in free energy (of the system) is opposite to the change in entropy of the universe.*** For example, if the system's free energy decreases by ten energy units there will be ten energy units expended to increase entropy in the universe — either in the system, in the surroundings, or in both combined.

Thus, the description of free energy solely in terms of system parameters does ***not*** mean that free energy is independent of the behavior of the surroundings. To the contrary, the ***Second Principle*** firmly links together the behaviors of system and surroundings by requiring that the ***sum*** of their entropies always increase. A decrease in entropy in one demands an increase in the entropy of the other, and any increase in the sum of their entropies — regardless of whether it originates in the system, the surroundings, or both — corresponds to a decrease in the (system) free energy.

We can now interpret equilibria in terms of free energy. A spontaneous process proceeds until the entropy of the universe is maximal, meaning that free energy of the system is minimal, because entropy of the universe and free energy of the system change in opposite directions. When no further entropy change occurs, no further free energy change occurs either — equilibrium has been reached.

A common interpretation of free energy is that it is energy available to do "useful" work, which immediately leads us to ask, "What is 'useless' work?" For our purposes "useless" work is associated with entropy increase and with volume change. We can thus think of free energy as what remains from the internal energy initially generated in a process after losses to entropy increase and to volume change are subtracted away (the latter explains why we added the assumption of "no volume change" three paragraphs back; it simplified things). For example, consider a source of chemical energy, like mitochondria oxidizing glucose to CO_2: from the maximum amount of energy available from the glucose, we lose some to entropy production and some (perhaps) to volume change. The remainder is free energy, which can be used usefully to move a muscle, to pump dissolved substance into a cell, to synthesize molecules, or to perform any other of the myriad tasks a cell requires.

The useful work done by the free energy will ultimately be reversed: the muscle will relax, the substance will diffuse back out of the cell, and the molecules will break up. These reversals will all be accompanied by an increase in the universe's entropy.

The free energy change in a reaction is obtained by subtracting the free energy of the initial state ***from*** that of the final state. Thus, a negative free energy change means that the free energy of the participants ***decreases*** during the process. Such ***exergonic*** processes are spontaneous because the entropy of the universe increases as system free energy decreases. On the other hand, free energy-requiring reactions are called ***endergonic*** and are not spontaneous. Because the free energy change depends only on the free energies of the initial and final states, we can conclude that the change is independent of the path or the sequence of intermediate states.

Path independence is illustrated by a ***free energy diagram,*** which schematically presents the free energy change of the system as a reaction, or group of reactions, progresses. For example, suppose reactant A is converted

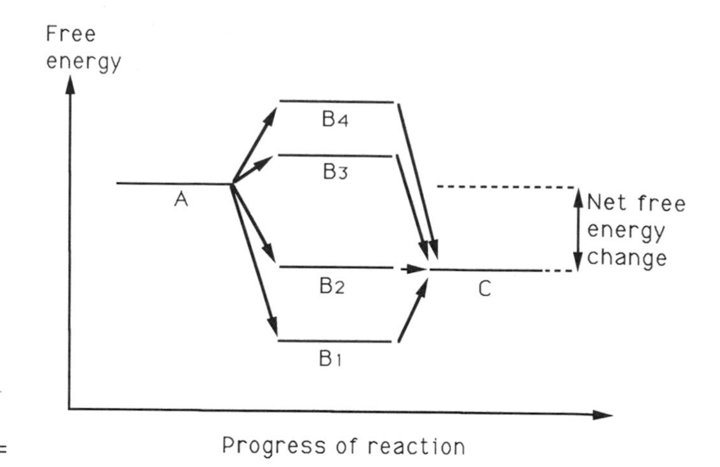

Figure 22.1
Four possible paths between states A and C. The net free energy change is the same for all four paths.

exergonically to product C and that the process involves an intermediate state B, as shown in Figure 22.1. Several paths between A and C are illustrated; the net free energy change is exactly the same in every case.

The "path-independence" of free energy is a very important behavior, one which is also exhibited by several other thermodynamic functions, e.g., entropy change. As shown in Figure 22.1, it allows a system to pass through any number of thermodynamically unfavorable intermediate processes, the only requirement being that the net transition from the initial to the final state be thermodynamically favorable (spontaneous).

Finally, the free energy change in a process is independent of the ***rate*** at which the process takes place. This is especially obvious when we compare the rate at which sucrose becomes CO_2, say, *in vivo* and *in vitro*. In the former case, conversions typically occur in fractions of a second and in the latter case the conversion of a like amount of sucrose may take months. The net free energy change is the same, however, in both cases. This should lead us to suspect that the cell has something extra to speed up the reaction *in vivo* (which, in fact, is the case and will be discussed in Chapter 24).

Applications, Further Discussion, and Additional Reading

1. An algebra-based tutorial on free energy is presented in *Guide to Cellular Energetics*, by Carter, L. C., W. H. Freeman, San Francisco, 1973. This book contains many numerical calculations of interest to biologists.
2. An extensive, calculus-based discussion of free energy can be found in *Physical Chemistry*, 3rd ed., by Atkins, P. W., W. H. Freeman, San Francisco, 1986.
3. Gibbs' free energy will reach a minimum at equilibrium in a system at constant temperature and pressure; under other conditions other parameters than free energy may be minimized. For example, suppose that the entropy and the volume of a system remain constant during a process. What then? Clearly the entropy of the universe will increase, but is there any *system* parameter, analogous to the free energy, that will decrease? In fact, the internal energy of the system performs that function. There is much more on this general topic in *Introduction to the Thermodynamics of Biological Processes*, by Jou, D. and Llebot, J. E., Prentice-Hall, Englewood Cliffs, NJ, 1990. (English translation by Cathy Flick.)
4. The book by Jou and Llebot (mentioned above in item 3) also contains a detailed discussion of nonequilibrium thermodynamics as it applies to biological systems. Nonequilibrium thermodynamics allows one to describe far-from-equilibrium situations quantitatively, which cannot be done by the usual methods of equilibrium thermodynamics.

Chapter 23
The Coupled-Reactions Model

An Exergonic Reaction can "Drive" an Endergonic Reaction

Two reactions are ***coupled*** if they share a common intermediate and if the overall free energy change is negative. Biological systems make extensive use of coupled reactions, as will be shown shortly. The reactions A → B and B → C are coupled if A → C is exergonic, B being the common intermediate, yielding A → B → C. It does not matter if either A → B or B → C is separately endergonic as long as A → C is exergonic overall because the path between A and C does not matter.

The coupled-reactions model has wide applicability. Several sets of coupled reactions are presented as examples in this chapter; in each case a free energy-releasing process drives a free energy-requiring one. Note that the three examples differ in the generality associated with the reactants and products.

Figure 23.1 shows one situation involving coupled reactions, where A, B′, B″, and C represent generic collections of compounds or atoms. As shown, A → B′ could represent solar thermonuclear reactions that produce light to drive photosynthesis and B″ → C could represent photosynthetic CO_2 fixation into

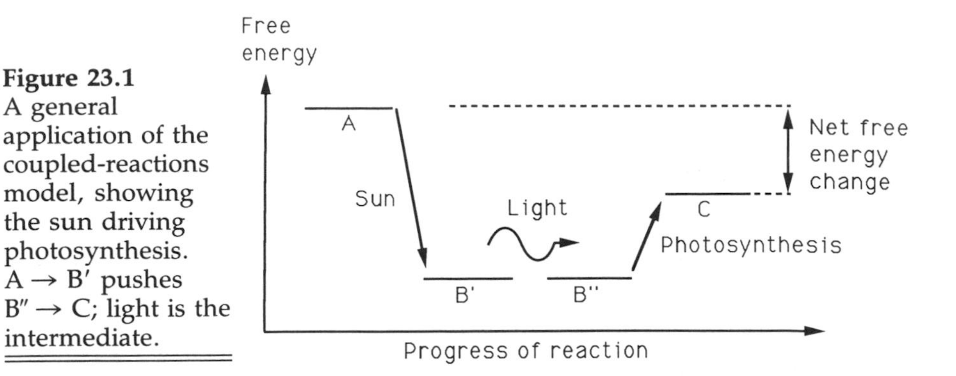

Figure 23.1
A general application of the coupled-reactions model, showing the sun driving photosynthesis. A → B′ pushes B″ → C; light is the intermediate.

glucose. The actual intermediate would be sunlight. Alternatively, A → B′ could represent the conversion of glucose to CO_2 and B″ → C could represent some ordering process, such as the polymerization of amino acids into a polypeptide. In that case, the actual intermediate would be adenosine triphosphate (ATP) formed in aerobic respiration and used in the polymerization.

By comparing Figure 21.7 and Figure 23.1, we can see the essential similarity between the flow of orderliness and the flow of free energy through a system. Free energy in sunlight or food is taken into the (local) living system; that free energy is then used to create order out of the nonliving precursors of which the living system is composed. After a number of such transfers the free energy is all lost.

As a second example of the coupled reactions model, we can be more specific about the natures of A, B′, B″, and C — in Figure 23.2 they are all relatively ***stable individual*** compounds. Here a pair of coupled reactions from glycolysis are shown. The free energy-releasing process of ATP hydrolysis drives, or pushes, the phosphorylation of fructose 6-phosphate to fructose 1,6-diphosphate. The actual intermediate is P_i, inorganic phosphate.

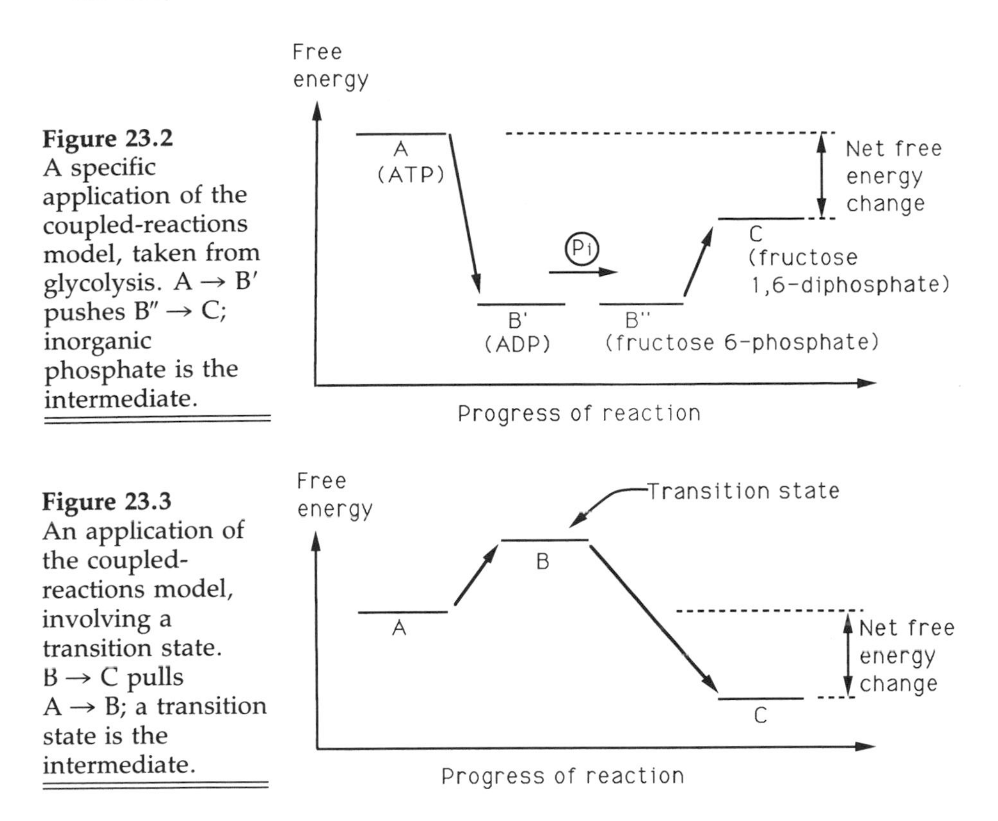

Figure 23.2
A specific application of the coupled-reactions model, taken from glycolysis. A → B′ pushes B″ → C; inorganic phosphate is the intermediate.

Figure 23.3
An application of the coupled-reactions model, involving a transition state. B → C pulls A → B; a transition state is the intermediate.

As a third example, in Figure 23.3, A and C are relatively stable compounds and B is a relatively unstable intermediate called a ***transition state***. (By "stable" we mean that a bottle of the compound could be stored on a shelf and used later. By "unstable" we mean that it has a lifetime less than, say, a second.) The exergonic reaction B → C pulls the coupled endergonic reaction A → B. We will discuss this situation in detail in the next chapter.

Applications, Further Discussion, and Additional Reading

1. There are many biochemical examples of the second of the applications given above. For examples see Chapters 11 through 19 of *Biochemistry,* 2nd ed., by Stryer, L., W. H. Freeman, San Francisco, 1981.

2. There is a discussion of coupled reactions in *Cellular Energetics,* by Carter, L. C., W. H. Freeman, San Francisco, 1973. This book, in tutorial form, contains many numerical calculations of interest to biologists.

Chapter 24
ACTIVATION ENERGY AND CATALYSIS

A TRANSITION STATE PREVENTS AN EXERGONIC REACTION FROM PROCEEDING

We now discuss the situation in Figure 23.3, where the intermediate state B is an unstable transition state between stable states A and C.

Suppose that A → C is an exergonic process; it should be spontaneous and there must be a reason if it has not yet taken place. For an illustration, look at Figurc 24.1; the process shown there will release free energy — but ***only*** if the energetic barrier shown leading to state C can be overcome.

The barrier, or transition state, in Figure 24.1 merely changes the path and therefore does not affect the net free energy change, but it clearly ***does*** obstruct the process from getting started in the first place. The free energy necessary to overcome such barriers is called the ***activation energy***; it is the free energy necessary to get a process going. As shown in Figure 24.1, the activation energy is invested in going from A to B and it is returned as the reaction goes over the barrier from B to C′. Any free energy obtained in going from C′ to C is net gain.

Figure 24.2 presents a macroscopic analog to an activation energy diagram; it shows a rock

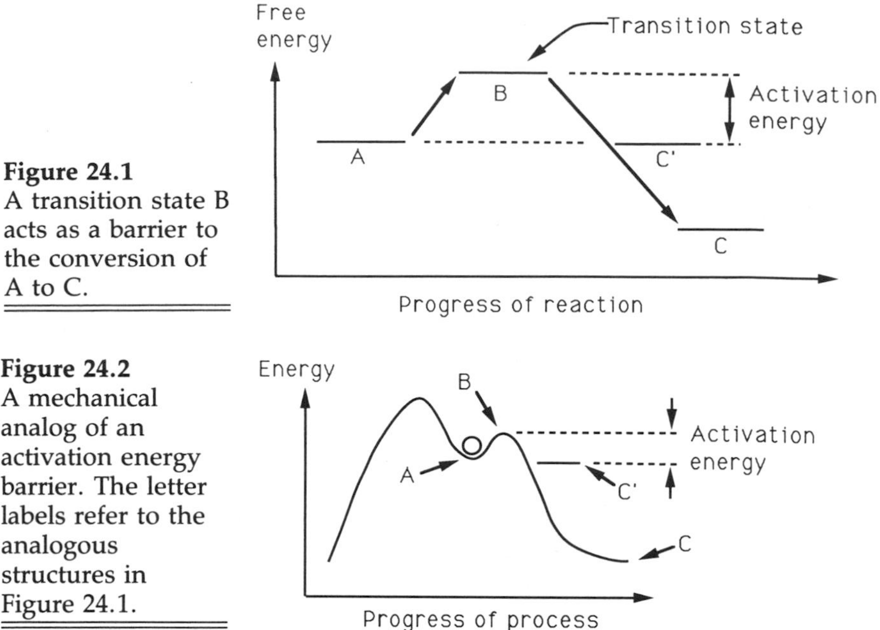

Figure 24.1
A transition state B acts as a barrier to the conversion of A to C.

Figure 24.2
A mechanical analog of an activation energy barrier. The letter labels refer to the analogous structures in Figure 24.1.

on a hillside. The spontaneous process of rolling downhill cannot begin until the downhill barrier is surmounted.

The statement two paragraphs back that "the activation energy . . . is returned" will be true no matter what the height of the transition state is as long as the overall process is exergonic. In this scheme we note that the reactions A → B and B → C are coupled via a common intermediate — the transition state B. Thus, B → C pulls A → B. The conversion of A to C evidently requires an initial uphill "push" into the transition state and the source of the required energy is of some interest.

Heat Can Provide Energy for Activation

We first inquire as to how *any* endergonic chemical reaction might generally be driven in the absence of being directly coupled to an exergonic one. Endergonic reactions of the cell usually require a few tenths of electron volts to move from stable reactants to stable

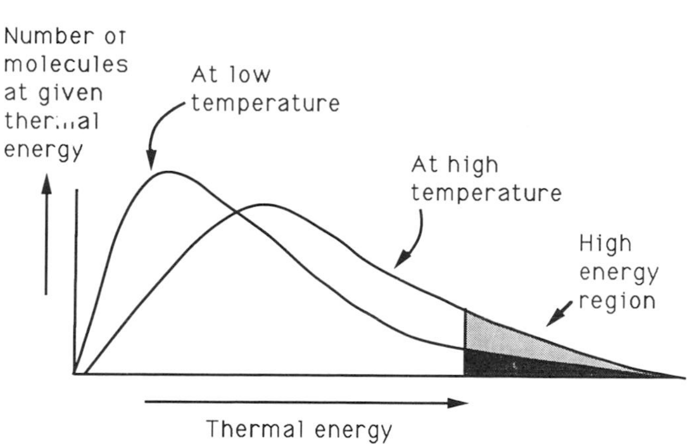

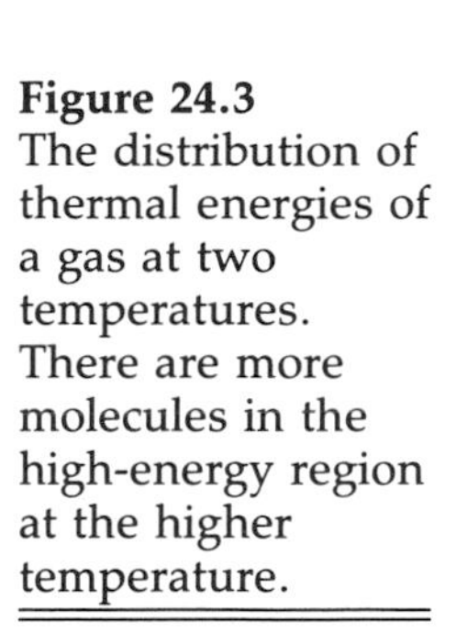
Figure 24.3
The distribution of thermal energies of a gas at two temperatures. There are more molecules in the high-energy region at the higher temperature.

products. Thus, as a first guess we could estimate that conversion to a transition state might require a few tenths of electron volts. The only energy available to an endergonic reaction in the absence of a coupled exergonic one, however, is heat from the surroundings of the reaction materials; at 30°C this amounts to about 0.02 eV/molecule. This thermal energy thus seems inadequate to drive endergonic reactions, which need at least a few tenths of electron volts.

There is a way out of this dilemma. When we say that thermal energy provides about 0.02 eV per molecule, that does not mean that ***every*** molecule has exactly 0.02 eV, merely that it is the average. In fact, from time to time a molecule will accumulate considerably more than the average thermal energy, pushing that molecule's energy close to what is needed to complete an endergonic reaction. This is illustrated in Figure 24.3, which shows the distribution of molecular energies in a collection of molecules at two different temperatures. The areas of the shaded regions show the number of molecules with relatively high energies for that temperature — perhaps enough to surmount the activation energy barrier.

Enzymes Reduce Activation Energy

Returning now to Figure 24.1, a typical biochemical activation energy may actually approach 1.0 eV! This is indeed a formidable barrier and few of the high-energy molecules of Figure 24.3 will ever accumulate this much thermal energy. Even if the overall reaction were exergonic it is clear that very little reactant would ever become product. In fact, if we concluded that the reaction did not go at all we would be close to the truth. Fortunately, this problem has a solution: the group of catalysts called ***enzymes*** reduce activation energies.

Virtually every one of the reactions in a cell is catalyzed by an enzyme specific for that reaction. In each case the enzyme lowers the activation energy to the point where available thermal energy is sufficient to allow the reactant (called the ***substrate***) to surmount the reduced activation energy barrier. It is not necessary that an enzyme reduce the activation energy to zero — merely reducing it from 1.0 to 0.3 eV can increase the reaction rate by 10^{12} times!

Figure 24.4 shows a reaction — with the transition state — in the presence and absence of an enzyme. We see that a rapid conversion of A to C can be expected in the presence of the enzyme but not in its absence. We also note that reduction of the barrier facilitates both the forward and backward reactions.

You should recall that, as shown in Figure 24.4, the free energy change during the conversion of A to C is independent of the level of the transition state B. In other words, the path, whether through B or B′, does not affect the free energy change. However, the path through B is much less probable than that

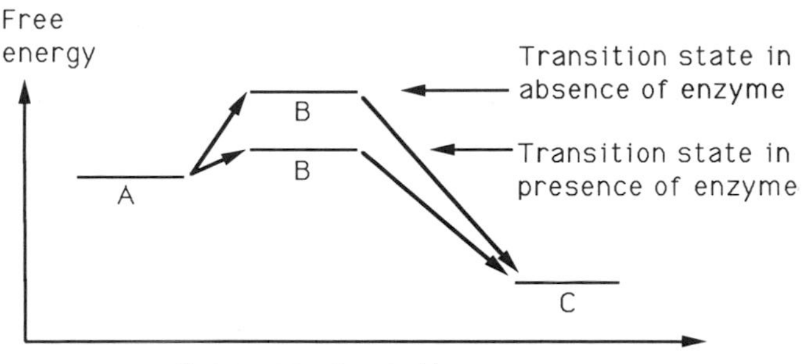

Figure 24.4
The effect of an enzyme on the activation energy. The reverse reaction is also affected, but it is not indicated by the arrows.

through B′, which is just a way of saying that the catalyzed reaction A → C can be expected to proceed rapidly and the uncatalyzed reaction will barely go.

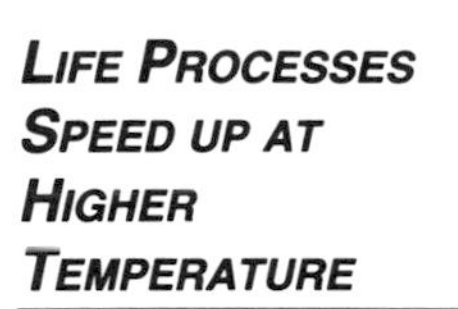

Life Processes Speed up at Higher Temperature

Enzymes lower the activation energy barriers that impede the initiation of exergonic reactions, but do not reduce them to zero. The remaining required activation energy can be provided by heat flow from the surroundings. Thus, the higher the temperature the more such free energy of activation is available and the faster the processes will go. The multiplication factor for the process rate per degree Celsius increase is called the Q_{10}. As a rough rule of thumb, the Q_{10} for most biological processes is about 2, meaning that a process whose rate is 1 at 10°C will have a rate of 2 at 20°C and a rate of 4 at 30°C. The "sunning" behavior of poikilotherms (cold-blooded animals) such as snakes and the spreading of the wings of perched insects such as butterflies are mechanisms to absorb sunlight to increase body temperature to the point where metabolic reactions can proceed at a biologically reasonable rate.

There is a limit to the temperature-generated increase of process rates in living systems: the various bonds that maintain a protein's tertiary structure typically dissociate somewhere around 45 to 60°C. Once that happens enzymes lose their catalytic function, reaction rates slow down dramatically, and the organism dies. Thus, desert reptiles seek shade at midday.

Applications, Further Discussion, and Additional Reading

1. The horizontal axis of Figure 24.3 is labeled, "thermal energy", a measure of the heat energy possessed by a particle. From the study of statistical mechanics we know that a manifestation of thermal energy is increased velocity; thus the axis could have also been labeled, "kinetic energy per particle". The latter makes good sense because those particles with high kinetic energy (in the shaded region) have the highest velocity and therefore the greatest chance of colliding with other reactants during a given time period. Such collisions have a good chance of resulting in a reaction.
2. We know from the ***Second Principle*** that no real process can be 100% efficient, but there is more that we can say about the origin of metabolic heat energy. The least amount of energy that a biological system can practically use from catabolic reactions, e.g., the oxidation of glucose, is that required to phosphorylate a molecule of adenosine diphosphate to adenosine triphosphate (about a third of an electron volt, depending on cellular conditions). Many exergonic chemical reactions in a cell release less energy than this, and such energy winds up as heat. The same considerations apply to anabolic reactions.

Chapter 25
Enzymes and the Determination of Cell Chemistry

Enzymes Control the Important Reactions in an Organism

The absence of an appropriate enzyme can reduce the rate of a reaction by a factor of 10^{-12}; we can thus say that the reaction essentially will not go without catalysis. This is not strictly true, of course, but is a good approximation. To this order of approximation, we can conclude that it is only the presence of the correct enzyme that determines whether or not a particular reaction goes forward at a biologically significant rate.

A human and an elephant have essentially the same reactants available to them — a mixture of gases (CO_2, etc.), minerals (metals, etc.), and biological compounds (glucose, amino acids, etc.). If every reaction were identical and proceeded equally fast in both the human and elephant, the two animals would have the same chemistry and would look alike (in fact, *be* alike). Suppose that the reaction $A \rightarrow C$ proceeded only in the presence of E_{AC}, an enzyme found only in humans. Suppose further that the reaction $A \rightarrow F$ proceeded only in the presence of E_{AF}, an enzyme found only in elephants. This is shown in Figure 25.1.

Figure 25.1
Two different enzymes catalyze two different reactions, starting with the same reactant.

C
E_{AC} (in humans)
A
E_{AF} (in elephants)
F

Two chemical paths are available to the reactant A, depending on whether enzyme E_{AC} or enzyme E_{AF} is present. This really means that A becomes C in humans, but becomes F in elephants.

We could now picture a whole sequence of enzyme-catalyzed reactions such as those of Figure 25.2, beginning with a common precursor A and developing into human-specific or elephant-specific products, as dictated by the presence of human-specific or elephant-specific enzymes.

We conclude that many important differences in organisms can be attributed to the enzymes present — not to specific reactants, the latter being generally available to any organism in the vicinity.

This line of reasoning can be extended to any two different organisms, e.g., humans and

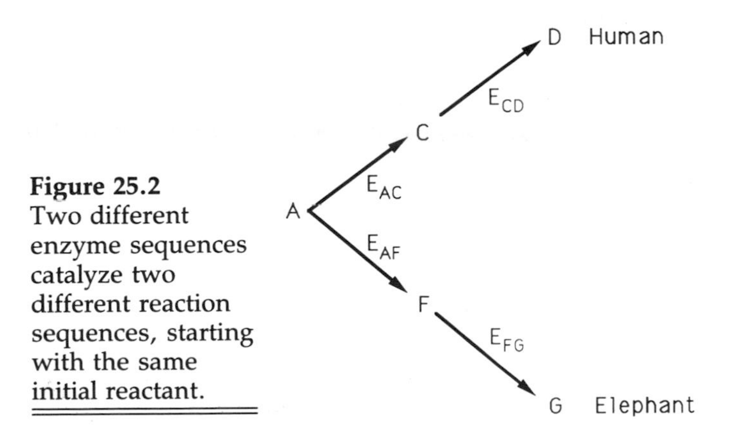

Figure 25.2
Two different enzyme sequences catalyze two different reaction sequences, starting with the same initial reactant.

rose bushes, or two different humans. The essential distinction between a dark-haired person and a fair-haired person is not in the food they eat or the air they breathe — those are shared in common; the difference is in the presence of enzymes that catalyze reactions leading to the formation of certain pigment molecules found in hair.

APPLICATIONS, FURTHER DISCUSSION, AND ADDITIONAL READING

1. The above discussion on the enzymatic determination of organismal chemistry is a conceptually simple one, resting entirely on chemical effects, and in a more realistic description we would have to include some other factors. For instance, two chemically identical organisms will probably be phenotypically different if they are placed in different environments; the expressions of many genes are sensitive to environmental factors. Further, two organisms might produce the same enzymes, but differ from one another in the amounts or in the cellular locations of the enzymes. Even with identical conditions of precursor supply the two organisms would be different. Of course, this argument immediately brings up the question, "Why are the amounts and locations different?" — the answer to which is that the organisms differ in ***other*** enzymes.

2. As an example of the scheme presented in Figure 25.1, consider the following: CO_2 is combined with water to give carbonic acid in our blood; the reaction is catalyzed by the enzyme carbonic anhydrase. On the other hand, in chloroplasts carbon dioxide is combined with ribulose bisphosphate to give a six-carbon compound; the reaction is catalyzed

by ribulose bisphosphate carboxylase. Surely carbon dioxide and water are in the chloroplast but, in the absence of carbonic anhydrase, essentially no carbonic acid is formed. It is the presence or absence of the relevant enzymes that determine the direction of the carbon dioxide's chemistry.

Chapter 26
MATERIAL TRANSPORT

PASSIVE DIFFUSION RESULTS FROM THE DRIVE TO GREATER ENTROPY

In the discussion of the ***Second Principle*** in Chapter 20, we represented a low entropy situation by 4 marbles restricted to 4 specific squares; we then represented a higher entropy situation by 4 marbles restricted to any 4 of 25 squares. Analogous physical situations are, respectively, a drop of ink localized to a small region in 1 L of H_2O and the same amount of ink diluted in 1 L of solution. The spontaneous progression of events will be from the former to the latter because the latter is more probable (there are more ways to obtain it).

The change in the distribution of the drop of ink, from concentrated to dilute, is predicted by the ***Second Principle*** — the orderly arrangement spontaneously progresses to the disorderly arrangement. This particular mechanical progression is called ***diffusion***, sometimes said to be ***passive*** to indicate its spontaneity. Passive diffusion is just a manifestation of the universe's irresistible tendency toward greater entropy, although the specific driving force behind the ink's motion is the thermally generated Brownian motion of the water. Of course, the Brownian motion of the water is

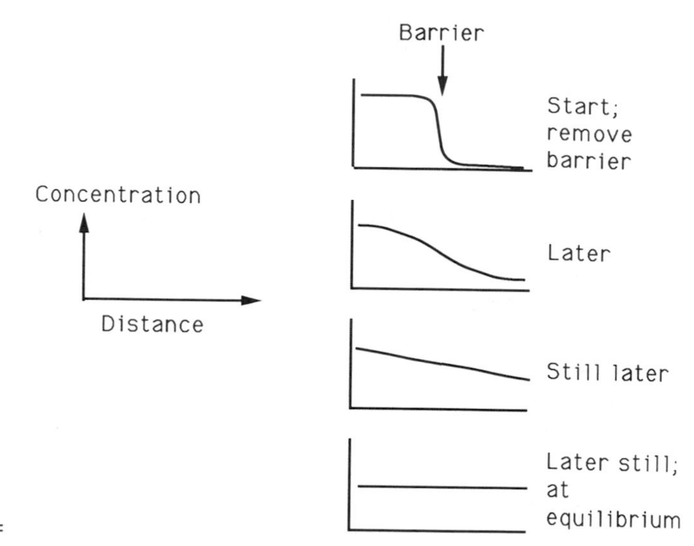

Figure 26.1
The change in the concentration of a solution with distance, over time.

itself a manifestation of the entropy induced by heat energy from the surroundings.

Diffusion Rates Depend on Area and Concentration Gradient

Fick's Law says that the net rate at which material passively diffuses from one region into another depends on two factors: the area of the boundary between the two regions and the extent to which the material's concentration differs between the two regions, i.e., the concentration gradient. We can make plausibility arguments for both of these factors. First, twice as much liquid should flow through a 2-cm^2 opening as through a 1-cm^2 opening. Second, there should be a net flow of material only when the material concentration is different in the two regions, the net flow always being from the region of high concentration to that of low concentration. Net flow ceases when the concentrations are the same, i.e., the concentration gradient is zero; to do otherwise could decrease entropy.

Figure 26.1 is a schematic description of material movement from one region to another.

Imagine that a thin barrier separates a concentrated solution from a zero-concentration solution. At time zero the barrier is removed and the concentrations in the two regions start to approach one another. Net material movement ceases when the two concentrations are the same, meaning that equilibrium has been reached.

Movement Against a Concentration Gradient Requires Energy

Cells are frequently known to accumulate dissolved substances in concentrations greater than those in the surroundings. In other words, entropy decreases in a local region; this process is called ***active transport***. As we have seen before, this is not forbidden by the ***Second Principle,*** it only means that entropy has increased more than that somewhere else in the surroundings.

Accumulating a substance against a concentration gradient reduces entropy in the system and thus is endergonic, requiring free energy. That free energy must come from an exergonic process — for example the hydrolysis of adenosine triphosphate (ATP) — which in turn drives some biological pump.

Active and Passive Transport in a Thylakoid

Hydrogen ions are pumped against a concentration gradient in a chloroplast thylakoid by using light energy as a free energy source. (A similar process occurs in mitochondria, driven by glucose metabolism.) This process can increase the hydrogen ion concentration by 10^{+3} inside the thylakoid, compared to its outside. The hydrogen ions can then spontaneously rush out via passive diffusion, driving the endergonic synthesis of ATP, as shown in Figure 26.2.

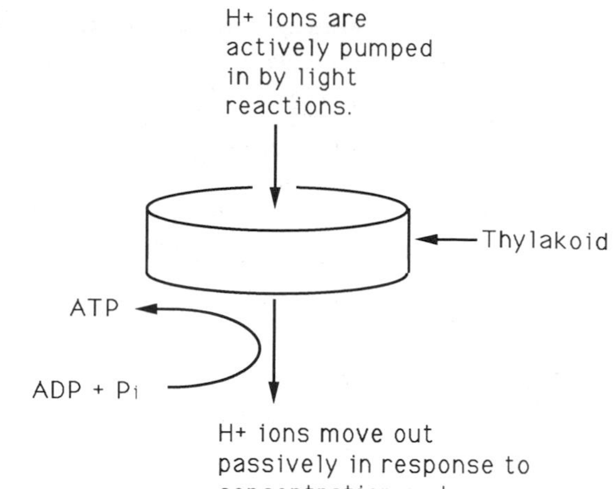

Figure 26.2
The movement of hydrogen ions in and out of a thylakoid.

Applications, Further Discussion, and Additional Reading

1. A good nonmathematical description of membrane physiology can be found in *Membranes and Their Cellular Functions*, by Finean, J. B., Coleman, R., and Michell, R. H., Blackwell Scientific, Oxford, 1984.

Chapter 27
METABOLIC HEAT GENERATION AND LOSS

WATER CAN STORE HEAT IN ORGANISMS

We have seen that the combination of kinetic energy associated with physiological temperatures and the enzymatic reduction of activation energy helps to make biological processes proceed efficiently. Physiological temperatures are usually above the temperature of the surroundings and thus require some immediate heat source, either internal or external.

Every real process generates heat somewhere in the universe, even the cooling that occurs inside a refrigerator causes heat production outside the appliance, as required by the ***Second Principle***. Thus, homeotherms (warm-blooded animals) can maintain their body temperature through the normal heat production associated with metabolism, the heat production being regulated by thermal sensors in blood vessels and the brain. They usually also have an external insulating layer of fat, feathers, or fur. Poikilotherms (cold-blooded animals) also produce heat through metabolism, but generally lack both insulation and neurological mechanisms for temperature regulation. Thus, they usually take the temperature of their surroundings unless they can absorb sunlight, as do many reptiles and insects for instance.

The energy of food is ultimately given up to heat, the temporary exception being that which is tied up at any given time in the orderliness associated with being alive, and ***that*** will eventually be lost to heat, too. Homeotherms, such as humans, must then maintain a balance between this generated heat and that lost (irreversibly) to their surroundings.

As pointed out in Chapter 21, we can imagine a flow of orderliness from the sun to green plants, to herbivores, and on through the food chain. Ordered structures pass their orderliness on down the chain with well under 100% efficiency, the differences spontaneously generating heat (and thus disorder). This heat can be temporarily "stored"; the water constituting the bulk of a living organism can hold onto a considerable amount of heat without the temperature increasing too much. Recall from Chapter 16 that liquid water is a network of hydrogen-bonded molecules. These hydrogen bonds (H bonds) can take up energy from the environment, vibrating with greater amplitude as they do. Thus, within reasonable limits the heat added to liquid water is rapidly diluted among the numerous H bonds, resulting in only a small temperature change. By the same token, the loss of heat does not lower the bulk temperature by much (again, assuming that the heat lost is not extreme).

This property of heat storage by a substance is measured by its ***heat capacity,*** which was discussed in Chapter 16 and whose units are calories per mass per degree. The heat capacity of water is 1 cal/(g · C): 1 cal added to 1 g of water raises the water's temperature by 1°C. By comparison, 1 cal added to 1 g of ethanol raises its temperature by almost 2°C; for most liquids the change is more severe still. Thus, one of the biologically useful properties of water is that it is a good thermostat, resisting temperature changes (upward and downward).

The "calories" used here are physical chemists' calories, not dietitian's calories. The latter are a thousand times the former. Thus, the statement that a spoonful of mayonnaise contains 100 (dietitian's) calories means that the mayonnaise, when burned, could raise the temperature of 10,000 g of water by 10°C!

The Effect of Volume and Surface Area on Metabolic Rate

The notion of heat "storage" by water does not imply a long-term effect. An organism eventually will lose heat to cooler surroundings and must produce or obtain more heat as compensation. For the case of homeotherms, we can make the approximation that the ***heat generated is proportional to body volume,*** i.e., doubling the volume doubles the mass, which results in twice as much metabolic heat generated. Further, ***heat loss to the surroundings should be proportional to surface area,*** i.e., doubling the area will provide twice the area through which heat loss can occur.

The effect of volume and surface area on heat production and loss can be seen in a simple model, shown in Figure 27.1. Two blocks of unit dimension represent organisms which have a volume of one unit and mass of one unit, the latter being a measure of heat production (double the mass and thereby double the amount of heat produced). Each organism loses heat through 6 exposed sides, for a total of 12. When pressed together, the total

Figure 27.1
The change in the surface-to-volume (S/V) ratio as blocks become stacked. The S/V ratio decreases as the blocks are stacked.

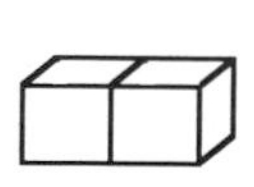

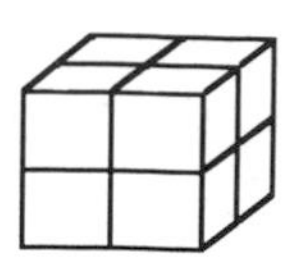

2 blocks
12 sides exposed
S/V = 12/2 = 6

2 blocks
10 sides exposed
S/V = 10/2 = 5

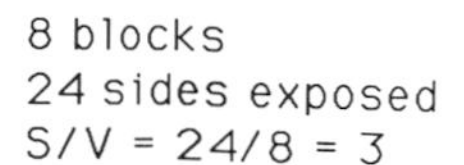

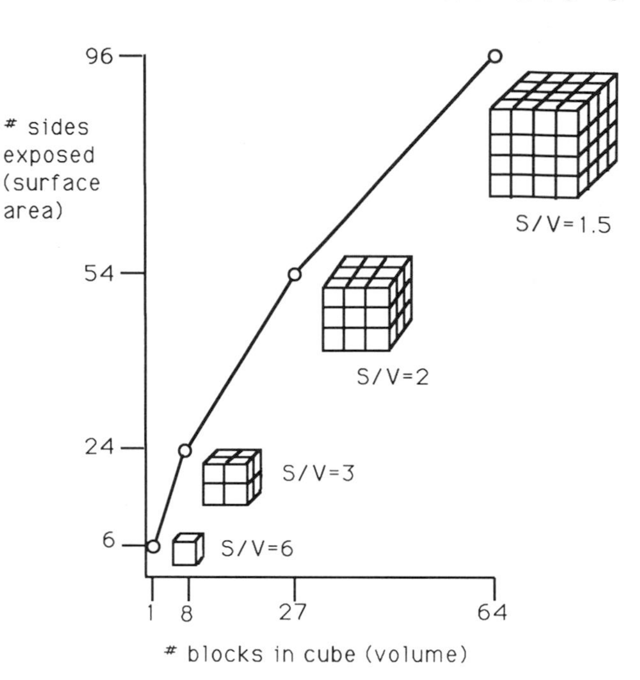

Figure 27.2
A graph of the surface area vs. the volume of a cube. The volume increases more rapidly than the surface area, resulting in a decreased S/V ratio.

(heat-producing) volume is still two, but the exposed (heat-losing) surface area is now only ten. A group of 8 blocks, arranged into a cube, has only 24 exposed sides, the other 24 covering one another in pairs. Notice how the surface-to-volume ratio decreases toward the right.

Figure 27.2 is a graph of exposed surface area vs. volume, for cubes of blocks; clearly the volume increases faster than the exposed surface area. Put another way, heat production increases faster than heat loss, as overall size increases. We would therefore expect a massive animal like an elephant to retain its heat better than a mouse would. This expectation is borne out by observation: an elephant has a lower basal metabolic rate than does a mouse, because the larger animal does not have to compensate for heat loss to the extent required by the smaller animal.

APPLICATIONS, FURTHER DISCUSSION AND ADDITIONAL READING

1. Chemical processes in a homeotherm normally take place under isothermal conditions. What heat sources and sinks maintain the temperature? You can find help with this question on pages 190 to 194 of *Biological Science*, 4th ed., Keeton, W. T. and Gould, J. L., W. W. Norton, New York, 1986.
2. The heat lost at the interface between an organism and its surroundings will depend on another factor besides surface area: the temperature gradient between the organism and its surroundings. The actual mechanisms by which heat is lost include evaporative cooling, infrared radiation emission, conduction (heating the surrounding air), and convection (passage of cool air or water past the skin).
3. The topic of heat generation and loss by organisms is discussed in the following two references:
   a. "Cold thermogenesis," by Alexander, G., in *International Review of Physiology*, Vol. 1 *(Environmental Physiology)*, Robertshaw, D., Ed., University Park Press, Baltimore, 1974.
   b. *Heat Transfer in Medicine and Biology*, Vol. 1, Shitzer, A. and Eberhart, R. C., Eds., Plenum Press, New York, 1985.

4. If a person has a mass of 80 kg, calculate how long he/she could live in a tight, perfectly insulated container before his/her body temperature reached a fatal level (42°C). Assume the person consumes 2000 dietician's cal/day (all of which produce heat) and that he/she is 100% water. (Answer: 5 h.)
5. Refer to Figure 27.1. Why do animals stay warmer when they huddle together?

Index

F

G

H

I

K

L

M

Q

R

S

X